The DNA Mystique

THE DNA MYSTIQUE

THE GENE AS A CULTURAL ICON

Dorothy Nelkin
M. Susan Lindee

W. H. Freeman and Company
New York

Library of Congress Cataloging-in-Publication Data

Nelkin, Dorothy.
   The DNA mystique : the gene as a cultural icon /
Dorothy Nelkin, M. Susan Lindee.
       p.   cm.
   Includes bibliographical references and indexes.
   ISBN 0-7167-2709-9
   1. Genetics—Folklore.   2. DNA—Folklore.
3. Genes—Folklore.   I. Lindee, M. Susan.   II. Title.
QH430.N46   1995
304.5—dc20   94-48711
                                                        CIP

© 1995 by W. H. Freeman and Company

Printed in the United States of America

1 2 3 4 5 6 7 8 9 0   VB   9 9 8 7 6 5

Contents

Preface		vii
1.	The Powers of the Gene	1
2.	The Eugenic Gene	19
3.	Sacred DNA	38
4.	The Molecular Family	58
5.	Elvis's DNA	79
6.	Creating Natural Distinctions	102
7.	Absolution: Locating Responsibility and Blame	127
8.	Genetic Essentialism Applied	149
9.	Genetic Futurism	169
10.	The Supergene	193
	Notes	200
	Sources	261
	Index	263

Preface

In the course of our work on the history of human genetics, on biological testing, and on the relationship between media images and public policy, we have repeatedly been struck by the powers ascribed to the gene in American culture.[1] At the same time, it became clear in our classes that college students' notions of heredity were frequently drawn from such sources as comic books, television sitcoms, science fiction, and other popular forums—sources in which the gene seemed to be daily increasing its authority and scope. We began to wonder what this meant for medical practice and for the social implications of the Human Genome Project. If the newly ascribed powers of the gene were taken seriously by consumers, would it shape their reactions to the institutional and political uses of genetic information? And what exactly did the gene seem to be capable of doing in the texts and images of popular culture? What powers did it possess? What problems did it seem to resolve? This book began from these questions, and these questions have guided it throughout. We have been studying the cultural construction of the gene by looking at a wide range of sources, and we have assumed that images that appear so consistently in so many unrelated forums reflect important cultural expectations.

As we quickly discovered, popular culture is, practically speaking, an infinite resource. Our notion of popular culture is broad. It includes textual and visual materials as well as the often fragmented elements of oral culture such as jokes, musical lyrics, radio talk shows, and clichés.[2] Our collection includes hundreds of stories and representations of the gene, not least ubiquitous images from the everyday detritus of soundbites, slogans, and snippets.

We have dealt with the problem of excessive material in four ways. First, whenever possible we availed ourselves of the databanks through which materials (in mass-circulation publications on genes and behavior, for example) could be retrieved. Second, we took advantage of the special knowledge of individuals who occupy particular niches of popular culture. We interviewed a sperm-bank founder, clerks in comic-book stores, journalists, geneticists, fertility specialists, and serious science-fiction and television fans. Third, we systematically reviewed the publications disseminated by special-interest groups including disability support foundations, infertility and adoption organizations, neo-Nazis, the religious right, and the population control movement. We read specialty publications such as *Soap Opera Digest;* and we canvassed child-care books, women's magazines, self-help books, and popular biographies for references to genes and heredity. Fourth, we attended discussions and debates on topics like the meaning of genetic definitions of homosexuality, the consequences of behavioral genetics for law and social policy, and the ethical, legal, and social implications of the Human Genome Project.

In these ways we attempted to sample, both broadly and strategically, a vast, quixotic data base. But perhaps of equal importance to our study was a less systematic network of informants—colleagues, family members, friends, and acquaintances—who heard of our project and immediately began pointing out examples and sending us jokes, clippings, movie recommendations, and copies of advertisements. We drew particularly on the special expertise of our students, well versed in popular culture, who often returned from their weekends with sightings (and citations).

This is not, therefore, a statistical study but an analysis of folklore. Our intention was to consider a broad range of sources and to explore many different forms of popular culture. Although we can show that images have changed through time, we have not compiled quantitative data and do not think it would have been appropriate or meaningful to do so. Nor is this a study that explicitly addresses ongoing debates in cultural studies about the relative passivity or activity of media audiences. We assume that the images and stories of genes in popular culture are not isolated artifacts but social products that both reflect and affect the cultural ethos; we assume, moreover, an active interaction between text and reader, media and audience. Indeed, we stress the diversity of people's interpretations and uses of the gene.[3] We show that many groups appropriate the gene to serve their needs—to promote their products, proclaim their solidarity, explain success or failure, deride or justify a cause—and we therefore construct popular culture as both an influence and a resource, a set of images to be accepted and to be reshaped to fit social agendas.

This kind of project relies on help from many people. The research assistance of Elizabeth Weinberg, a graduate student at New York University and an amazing connoisseur of popular culture, was especially helpful in collecting and keeping track of a large quantity of material. Other graduate students who helped collect material in specific areas included Mark Hamel, Betsy Hanson, Margo Hoffman, Lissa Hunt, Douglas Keen, Rosann Kosta, Ellen Kruger, Howard Lune, Neda Pourang, and Kevin Tucker. Many colleagues helped in providing material and critiquing drafts: Jon Beckwith, Paul Billings, Jo Dixon, Rochelle Dreyfuss, Chris Feudtner, Sarah Franklin, David Garland, Laura Lee Hall, Elizabeth Knoll, Mark Nelkin, Robert Proctor, Rayna Rapp, Charles Rosenberg, Brett Skakun, Ann Snitow, Christina Spellman, Sherry Turkle, and Keith Wailoo. Our editor, Jonathan Cobb, was a relentless and valuable critic. Christine Hastings cheerfully guided our manuscript into print.

PREFACE

Program officers at the ELSI (Ethical, Legal and Social Implications) Program of the Center for Human Genome Research of the National Institutes of Health were very helpful. Eric Juengst, Elizabeth Thomson, and Betty Graham sent comic strips and cartoons.

We wish to acknowledge research support from the National Center for Human Genome Research of the National Institutes of Health, Grant 1RO1 HG0047-01, and from our respective institutions, New York University and the University of Pennsylvania.

Dorothy Nelkin
Susan Lindee
December 1994

1

The Powers of the Gene

A full-color advertisement boasts that a new BMW sedan has a "genetic advantage"—a "heritage" that comes from its "genealogy."[1] A *U.S. News and World Report* article on the Baby M custody dispute states that it will not make much difference which family brings up the child since her personality is already determined by her genes.[2] A cartoonist lists genetically linked traits: "excessive use of hair spray, bottomless appetite for country-western music, right wing politics."[3] A critic reviewing a play about the persistence of racism says, "it's as if it has a DNA of its own."[4] The term "gene pool" appears as the name of a comedy group, the title of a TV show, and in the captions of comic books and the lyrics of rock music. Indeed, in the 1990s "gene talk" has entered the vernacular as a subject for drama, a source of humor, and an explanation of human behavior.[5]

In supermarket tabloids and soap operas, in television sitcoms and talk shows, in women's magazines and parenting advice books, genes appear to explain obesity, criminality, shyness, directional ability, intelligence, political leanings, and preferred styles of dressing. There are selfish genes, pleasure-seeking genes, violence genes, celebrity genes, gay genes, couch-potato genes, depression genes, genes for genius, genes for saving, and even genes for sinning. These popular images convey a striking picture of the gene as powerful, deterministic, and central to an understanding of both everyday behavior and the "secret of life."[6]

What is this crucial entity? In one sense, the gene is a biological structure, the unit of heredity, a sequence of deoxyribonucleic acid (DNA) that, by specifying the composition of a protein, carries information that helps to form living cells and tissues.[7] But it has also become a cultural icon, a symbol, almost a magical force. The biological gene— a nuclear structure shaped like a twisted ladder—has a cultural meaning independent of its precise biological properties. Both a scientific concept and a powerful social symbol, the gene has many powers.

In this book we explore those powers, showing how the images and narratives of the gene in popular culture reflect and convey a message we will call genetic essentialism.[8] Genetic essentialism reduces the self to a molecular entity, equating human beings, in all their social, historical, and moral complexity, with their genes.

DNA in popular culture functions, in many respects, as a secular equivalent of the Christian soul. Independent of the body, DNA appears to be immortal. Fundamental to identity, DNA seems to explain individual differences, moral order, and human fate. Incapable of deceiving, DNA seems to be the locus of the true self, therefore relevant to the problems of personal authenticity posed by a culture in which the "fashioned self" is the body manipulated and adorned with the intent to mislead.[9] In many popular narratives, individual characteristics and the social order both seem to be direct transcriptions of a powerful, magical, and even sacred entity, DNA.

Increasing popular acceptance of genetic explanations and the proliferation of genetic images reflects, in part, highly publicized research in the science of genetics. Such research, however, occurs in a specific cultural context, one in which heredity and natural ability have often seemed important to formulations of social policy and social practice.[10] Old ideas have been given new life at a time when individual identity, family connections, and social cohesion seem threatened and the social contract appears in disarray.

Changing technologies for the manipulation and assessment of DNA have, moreover, dramatically changed the social implications of these revived ideas. It seems imperative, therefore, to examine these trends critically at a time when diagnosis and prediction based on DNA analysis have so many new applications. In the laboratory, DNA can be used to detect unseen conditions of risk and predict future conditions of disability or disease. Within the family, DNA can be used to define meaningful relationships and make reproductive decisions. In the larger culture, DNA can be used to locate responsibility and culpability, as well as to justify social and institutional policies. Those on all sides of the political spectrum can proclaim that specific biological properties of DNA lend support to their policies or goals. And their claims all build on the DNA mystique.

Yet in the history of biology there are few concepts more problematic than that of the gene. It began as a linguistic fiction, coined by Danish geneticist Wilhelm Johannsen in 1909 to describe a presumed cellular entity capable of producing a particular trait. He drew the term from German physiologist and geneticist Hugo DeVries's "pangenes," a term derived from Charles Darwin's "pangenesis," a theory of the origins of biological variation. For the first generation of experimental geneticists (in the early twentieth century), a "gene" was, in practice, a physical trait—the wing shape or eye color of the fruit fly *Drosophila*, for example—which seemed to derive from a substrate of hereditary material, the actual constitution and functioning of which were unknown at the time.[11]

3

In the post–World War II era, the increasing elucidation of the gene as a molecular entity has both clarified its physical form—a double helix of deoxyribonucleic acid (DNA)—and complicated its biological meaning.[12] As contemporary genomics science has demonstrated, DNA does not produce bodily traits in a simple, linear way. It interacts with itself and with its larger environment: Identical sequences of DNA in different locations on the genome (the entire complement of DNA in any given organism's cells) can have different biological meanings. And different genes can have identical effects in different people. Genomes also have large regions, so-called "junk DNA," that seem to have no function at all.

For contemporary molecular geneticists, "gene" is convenient shorthand, referring generally to a stretch of DNA that codes for a protein. In the sense that some sections of DNA produce specific biological events, genes are real entities, but their workings are not simple. While increasing scientific knowledge of molecular processes has clarified some questions, it has also raised new and unexpected ones. Why is so much of the genome without obvious function? Why do many genetic diseases become more severe from generation to generation as a consequence of reduplication of short coding sections? And what does the ambiguity of the genome—its biological indeterminism—mean for our understanding of evolution and evolutionary processes? Much of this complexity disappears when the gene serves its public roles as a resource for scientists seeking public support and as a popular explanation for social problems and human behavior, and a justification for policy agendas.

The point of our analysis is not to identify popular distortions of science or to debunk scientific myths. The interesting question is not the contrast between scientific and popular culture; it is how they intersect to shape the cultural meaning of the gene. Some of the images we explore draw on well-established scientific ideas, some on findings that geneticists continue to question, while others seem to be independent of biological research. The precise scientific legitimacy of any image, however, is less important than the cultural use that is made of it. How do scientific concepts serve social ideologies and institutional agendas? Why do certain concepts gain social power to become the focus of significant popular and scientific attention? And what role do scientists play in shaping the appropriation of such concepts?

Science and Culture

It is not a coincidence that the popular appropriation of genetics has intensified just as scientists around the world have begun an effort to map and sequence the entire human genome, for in presenting their research to the public, scientists have been active players in constructing the powers of the gene. The cutting edge of this scientific exploration is the Human Genome Project, an international scientific program to map and sequence not only the genes but also the noncoding regions of all the DNA contained in the 24 human chromosomes.

Although gene mapping began in the 1910s with studies of the common fruit fly, *Drosophila*, large-scale mapping of human genes was not technically feasible until the development of greater computer capacity and a variety of new laboratory techniques in the 1980s. Building on these techniques, the human gene mapping program began in 1989 in the United States, where it is funded through the National Institutes of Health (NIH) and the United States Department of Energy (DOE) at a total anticipated cost of more than $3 billion over 15 years. Similar projects are underway in Great Britain, Japan, Russia, the European Community, and other industrialized nations.

Genome researchers hope to locate and determine the exact order of the base pairs in the estimated 100,000 human genes, as well as in the many sections of DNA with no known function. As of the summer of 1994 geneticists had identified over 2500 of the 3000 genetic markers (sections of DNA that can be used as signposts along the genome) needed to create a genetic map. Many single gene disorders—diseases caused by a known form of a particular gene—are already located on the map, either directly or through the identification of genetic markers that "follow" the disease through large, well-characterized family groups. These included cystic fibrosis, retinitis pigmentosa, one form of Alzheimer disease, and more rare conditions such as Huntington's disease, Gaucher disease, malignant hyperthermia, and epidermolysis bullosa.[13]

Geneticists are also exploring the patterns of inheritance of conditions with apparent multiple gene involvement, suggesting familial predispositions to some forms of cancer and Alzheimer's, emphysema, juvenile diabetes, cleft palate, heart disease, and mental illness. Researchers have identified the genetic markers for certain kinds of breast and ovarian cancer and have located the gene causing the mutation responsible for some colon cancers. One goal of such research is to identify susceptible individuals before their symptoms appear.[14]

Seeking to assure continued public funding of a long-term, costly project, genome researchers have been writing for popular magazines, giving public talks, and promoting their research in media interviews. They contribute to popular imagery as they popularize their work in ways that resonate with larger social concerns. Indeed, many of the values and assumptions expressed in popular representations of genes and DNA draw support from the rhetorical strategies of scientists—the promises they generate and the language they use to enhance their public image.[15]

Three related themes underly the metaphors geneticists and other biologists use to describe work on the human genome. These are a characterization of the gene as the essence of identity, a promise that genetic research will enhance prediction of human behavior and health, and an image of the genome as a text that will define a natural order.

Some scientists borrow their images from the computer sciences: The body is less a conscious being than a set of "instructions," a "program" transmitted from one generation to the next. People are "readouts" of their genes. If scientists can decipher and decode the text, classify the markers on the map, and read the instructions, so the argument goes, they will be able to reconstruct the essence of human beings, unlocking the key to human ailments and even to human nature—providing ultimate answers to the injunction "know thyself." Geneticist Walter Gilbert introduces his public lectures on gene sequencing by pulling a compact disk from his pocket and announcing to his audience: "This is you."[16]

Other metaphors used by scientists imply the possibilities of prediction, encouraging the use of their science for social policy. They call the genome a "Delphic oracle," a "time machine," a "trip into the future," a "medical crystal ball." Nobelist and first director of the U.S. Human Genome Project James Watson says in public interviews that "our fate is in our genes."[17] Futuristic scenarios promise that genetic prediction will enhance control over behavior and disease. Thus, a geneticist promises that "present methods of treating depression will seem as crude as former pneumonia treatments seem now."[18] A food scientist writes that food companies will sell specialized breakfast cereals to consumer targets who are genetically predisposed to particular diseases. "Computer models in the home will provide consumers with a diet customized to fit their genetic individuality, which will have been predetermined by simple diagnostic tests."[19] And a biologist and science editor, describing acts of violence, editorializes that "when we can accurately predict future behavior, we may be able to prevent the damage."[20]

Scientific illustrations, too, glamorize DNA and promote the notion of genetic essentialism. The logo for the joint NIH-DOE publication, *Human Genome News*, portrays a human figure in silhouette, standing inside two swirling ribbons of DNA, contained within a circle. Enscribed around him are the names of scientific disciplines: "Chemistry, Biology, Physics, Mathematics, Engineering." The twisted double helix of DNA surrounding the figure suggests the imprisonment of the human being, who will be released through scientific knowledge. This logo conveys the power of science and its promise for the future.

Geneticists also refer to the genome as the Bible, the Holy Grail, and the Book of Man. Explicit religious metaphors suggest that the genome—when mapped and sequenced—will be a powerful guide to moral order. Other common references to the genome as a dictionary, a library, a recipe, a map, or a blueprint construct DNA as a comprehensive and unbiased resource, an orderly reference work. The population geneticist Bruce Wallace has compared the human genome to "the torn pages of a giant novel, written in an unknown language, blowing about helter-skelter in an air-conditioned, enclosed space such as Houston's Astrodome."[21] Wallace's chaotic image of the genome implies the promise that scientists engaged in mapping the human genome will (eventually) capture all the pages, put them in proper order, translate the language, and analyze the meaning of the resulting text.

The apparent precision of a map may make invisible the priorities and interests that shaped it. As forms of knowledge, all maps reflect social perspectives on the world at the time of their making; they are the products of cultural choices. Maps select and link features of the world, in effect transforming those features by making them part of a coherent, single landscape. The selectivity of maps is a part of their visual power, of course, for they are also instruments of persuasion. As one curator put it, "Every map is someone's way of getting you to look at the world in his or her way."[22] Map imagery suggests that once a gene is located, its interpretation will be objective and independent of context. But as molecular biologist Christopher Wills has observed, "simply determining the sequence of all this DNA will not mean we have learned everything there is to know about human beings, any more than looking up the sequence of notes in a Beethoven sonata gives us the capacity to play it."[23] A mapped gene may appear to be a straightforward detail, to be extracted and understood without reference to culture and experience. Yet the language of the genome, like the language of a dictionary, must be contextualized to be understood. Genes are, like words, products of (evolutionary) history, dependent on context, and often ambiguous, open to more than one interpretation.[24]

Meanwhile, the successes of molecular genetics and the high profile of the Human Genome Project are shaping the assumptions underlying research in other scientific fields. Behavioral geneticists and psychologists, working with human twins and extrapolating from animal models, have attributed shyness, intelligence, criminality, even religiosity and other complex human traits to heredity. The Minnesota Center for Twin and Adoption Research has provided percentage estimates of the extent to which certain personality traits are determined by heredity: extroversion, 61 percent; conformity, 60 percent; tendency to worry, 55 percent; creativity, 55 percent; aggressiveness, 48 percent.[25] While human genome research has been promoted as a way to find disease genes, many within the scientific community believe that a map of the genome will also document the inheritance of these complex, socially important human traits. Indeed, some scientists believe this is a major goal. Nobelist David Baltimore has commented that the genome project "will allow us to examine human variability, for example, variations in mathematical ability, or what we call intelligence. . . . The rationale is not to find human disease genes, because we're doing moderately well at finding them right now. But the only way to study the genetics of the higher perceptual, higher integrative human functions is by actually studying human beings. . . . The genetic and physical maps are designed for that."[26]

The emphasis on genes for specific behaviors, however, is controversial. Some scientists argue that efforts to measure the relative effects of heredity and environment on behavior systematically misconstrue the two as independent rather than interactive forces, underestimating the influence of environmental forces on gene expression.[27] Critics point out that the heritability of any trait is simply a statistical construct that may suggest variations between populations but may have no simple meaning for the individual.[28] Stephen Jay Gould has observed that efforts to distinguish the relative effects of nature and nurture propose a false dichotomy by confusing correlation with causation: "Genes influence many aspects of human behavior, but we cannot say that such behavior is caused by genes in any direct way. We cannot even claim that a given behavior is, say, 40% genetic and 60% environmental. . . . Genes and environment interact in a non-additive way."[29]

Some critics question the motivation behind efforts to measure the relative effects of nature and nurture on behavior. Psychologist Douglas Wahlsten, for example, believes that "the only practical application of the heritability coefficient is to predict the results of a program of selective breeding."[30] And African American organizations, sensitive to the racist implications of deterministic explanations of deviance, attacked plans for a scholarly conference on "genetic factors in crime."[31]

Despite continued controversy over methods and motives, efforts to determine the genetic basis of human behaviors such as alcoholism and crime draw legitimacy from the rising fortunes of molecular biology. These efforts have captured public attention, for such research addresses critical social questions—about the basis of human identity and individual differences, the nature of deviance, and the location of responsibility for social problems.

Scientists often dismiss as oversimplified and distorted the way their work is appropriated. But the relationship between scientific and public culture is far more complex. As historian Robert Young put it, it is often "impossible to distinguish hard science from its economic and political context and from the generalizations which serve both as motives for the research and which are fed back into social and political debate."[32]

The history of science is the story of the selective analysis of reality, and many of the most interesting problems raised by historians and sociologists focus on this selectivity; that is, on how science, as a human and cultural process, can both depict nature and create culturally specific knowledge. As recent social and historical studies of science have suggested, an observation becomes a fact through social negotiation of critical questions. What does an experiment mean? What can it prove? What counts as a demonstration of a particular phenomenon? And what counts as a satisfying explanation? A different explanation, equally consistent with the phenomena observed, may not be accepted if it violates the researcher's expectations.[33] These expectations, often unarticulated, can be difficult to recognize, so integral are they to the culture of science.[34] And scientific culture itself reflects larger cultural concerns and assumptions—a point that seems obvious when we look back to the seventeenth century, but less evident when we consider contemporary science.

Popular imagery is far more transparent than scientific discourse. It provides a way to gain access to the social concerns and common understandings that are shaping contemporary concepts in both the science of molecular genetics and the culture at large.[35]

The Importance of Popular Culture

Popular culture matters. For many consumers, media stories, soap operas, advice books, advertising images, and other vehicles of popular culture are a crucial source of guidance and information. These are not simply escapist sources. They are narratives of meaning, helping their attentive listeners deal with social dilemmas, discover the boundaries of socially acceptable behavior, and filter complex ideas. The stories in this literature bear on personal decisions and determine the acceptability of social and institutional policies.

Mass culture, writes critic Robert Warshaw, is "the screen through which we see reality and the mirror in which we see ourselves."[36] Popular narratives of television and pulp fiction convey theories of human motives and representations of social relationships. They employ conventions that are readily recognized by their audience and that conform to and shape cultural expectations and the emotional structure of everyday beliefs. "Habitual images and familiar metaphors," says anthropologist Marilyn Strathern, "provide the cultural forms that make ideas communicable."[37] Some ideas, writes legal scholar Peggy Cooper Davis, are "proverbial stories" that become, through repetition, so familiar they form the "unconscious habits of thought," the unarticulated, taken-for-granted assumptions and patterns of belief that shape our concepts of social order.[38] The study of popular culture offers a way to examine the changing constellations of meaning in a society and to understand its shared assumptions.[39]

Media images do not necessarily determine individual behavior; the effect of popular images on individual choices is tempered by prior attitudes, expectations, and experiences.[40] But persistent images reveal the kinds of behaviors that are culturally valued, creating a framework for people's expectations. Historian Allen Batteau has observed that "texts provide an archetypal reality and hence the elementary structure of [the] collective conscience" of different groups.[41] Similarly, sociologist Todd Gitlin argues that the media, through repeated stories, can create a "hegemonic ideology" that defines certain actions and thoughts as "natural."[42] And more specifically, in interviews with people in genetic counseling clinics, anthropologist Rayna Rapp found that responses to genetic information were directly influenced by the world of "medicalized melodramas"—"Dallas," "St. Elsewhere," and the "Jerry Lewis Telethon."[43]

Repeated metaphors also serve to define experience, cultivate stereotypes, and construct shared meanings. George Lakoff and Mark Johnson observe that metaphors affect how we perceive, think, and act. Metaphors structure our understanding of events, convey emotions and attitudes, and allow us to place public issues and events in a shared context of common belief.[44]

The images that appear in commercial mass culture are especially revealing of such common beliefs. When a catalog for public television fans offers a "double helix bracelet" that "looks great with genes," such a promotion reflects a calculated evaluation of demographics.[45] Television stories are especially important in light of their mass audience; the average American watches television 7.5 hours each day. Those planning this programming cater to popular tastes in their struggle to improve ratings. Advertising images are significant indicators of public perception and belief, for they are the products of deliberate marketing strategies based on research about what will sell. When advertisements for automobiles, cosmetics, and even sneakers use genetic language and images ("Thanks for the genes, Dad,"—in a blue jeans ad), it is because their creators believe that this message will be broadly understood and widely appreciated. Popular culture is a business, and the appeal of any product—a television sitcom, advertising campaign, or newspaper story—depends on its resonance with consumer experience and popular belief.[46]

In his novel *Immortality*, Milan Kundera invented the word "imageology" to capture an important phenomenon of the late twentieth century. "We can rightfully talk of a gradual, general, planetary transformation of ideology into imageology. . . . This word finally lets us put under one roof something that goes by so many names: advertising agencies; political campaign managers; designers who devise the shape of everything from cars to gym equipment; fashion stylists; barbers; show business stars dictating the norms of physical beauty. . . . Imagologues create systems of ideals and anti-ideals, systems of short duration that are quickly replaced by other systems but that influence our behavior, our political opinions and aesthetic tastes."[47]

Accordingly genetic images, found in many sources, appeal to television scriptwriters, advertising executives, journalists, and publicists. This diversity suggests that they are more than just the product of individual imaginations. Systemic images, repeatedly invoked in familiar and accepted patterns, cannot be explained solely by looking at the intention of the individuals who produced them. They are social products. Nor are images simply passive reflectors of values. The "imagologues," those agents of popular culture in the print and broadcast media, are also "architects of ideology," building and shaping social visions.[48] Through their images and stories, they create the frames that generate meaning by making the world intelligible to their readers or viewers.[49] Through their stories, they provide models of appropriate and acceptable behavior, shaping beliefs about society, and promoting ideas and expectations that define a communal culture.

Genetic images, of course, are not the only images present in popular culture, nor do they dominate the media in any statistical sense. At least as prevalent are accounts emphasizing the influence of environment and experience on behavior, and these can be equally deterministic, as childhood experiences of abuse or deprivation, for example, appear to have an inexorable effect manifested in utterly predictable syndromes in adult victims. Many media stories attribute adult behavior to personal experiences, whether impoverishment, trauma, or a supportive family environment. And in some popular texts, the human condition is a consequence of alien plots, past lives, demonic spirits, or divine intervention. But the gene today is a common and increasingly powerful presence in American popular culture. Unlike demons or past lives, moreover, DNA is the focus of a major international scientific program, the Human Genome Project, and this lends credibility to genetic explanations of an ever-expanding set of human behaviors. When Leo Buscaglia, the prolific, best-selling author and frequent TV guest expert on human relationships, wrote about love in 1972, he defined it as "a learned emotional reaction." In 1992, he began his book *Born for Love* by declaring that: "Love is certainly genetically inscribed."[50] In the 1990s, the gene has conquered new territories and gained new powers.

The Gene as an Icon

The "First Interstate Sperm Bank" was the setting for a 1991 episode of the popular prime-time television comedy series, "In Living Color."[51] As the scene opened, a doctor at the sperm bank handed a customer a small plastic container and thanked her for her patronage. Suddenly an armed woman barged in demanding to be given the sperm of Denzel Washington. The bank officer explained that the popular actor's sperm was not available, but offered instead that of athletes Mike Tyson or James Worthy. The agitated "bank robber" shouted that both had the "wrong genetics." Desperate to have a baby because her "biological clock is ticking," she accepted a stranger's offer to solve her dilemma the old-fashioned way. As she left, however, she shot a container of sperm labeled "The New Kids," announcing "I just saved everyone from another generation of 'New Kids.'"

The humor in this television episode drew on ideas about genes, heredity, and reproduction familiar to a prime-time television audience. The violent confrontation over sperm—presented as a sitcom satire—mocks the idea that personality and behavior are inherited and caricatures the desperation of those whose "biological clocks" inspire a sudden desire to become mothers. It also parodies concerns about the threat of future genetic decline through the reproduction of the unfit—in this case, the "unfit" being the (all white) members of the highly promoted singing group "New Kids on the Block," viewed by some in the African American community as rap music rip-off artists with degenerate musical taste.

This narrative of the genetic origins of behavior and the threat of future genetic decline conveys a cluster of expectations and fears that have significant historical resonance. Long before heredity could be biologically explained, notions of "blood" and kinship were used to account for social inequalities. The superiority of "blue bloods," the basic evil of "bad seeds," and the idea that "blood is thicker than water" were important historical themes, reflecting social interest in family relationships.[52] Legends of baby switching, the Oliver Twist story, and the folk tale of the "incognito prince" suggested the overwhelming power of "blood ties" to determine a child's fate.[53] Heredity has long been interpreted as socially powerful, though the relative importance of heredity and environment has been the focus of a continuing debate.

Renewed interest in genetic explanations reflects the high status of the science of molecular genetics. But it is also a

15

response to the stresses and strains of an increasingly secular-ized, complex, and seemingly chaotic society. Biological expla-nations often appear to be more objective and less ambiguous than environmental or social ones. They also promise biomed-ical control of social problems. At a time of concern with ethnic and class differences, genetics as a science of dif-ferences seems to provide reliable, clear-cut ways to justify social policies on the basis of "natural" or predetermined characteristics, to differentiate "them" from "us." At a time of significant public concern about alcoholism and crime, genet-ics as an explanation of good and evil seems to provide hard and certain ways to codify what is normal or deviant.

DNA, to borrow from Sherry Turkle's analysis of Freudian and computer concepts, has become "an object to think with," a malleable idea by means of which different interpre-tive communities can express diverse, even contradictory, concerns.[54] Genetic metaphors are used to buttress class dif-ferences (the result of "good breeding") and to reinforce social stereotypes ("differences lie in the genes"). They serve to explain human exceptionalism on the basis of different DNA ("the genes of genius"), but also to claim the rights of animals on the basis of shared DNA ("A rat is a pig is a dog is a boy"[55]). Genetics can justify social harmony (based on common ancestry) or social divisions (based on race). Genetic explanations can absolve an individual from respon-sibility for action, but the individual's genes can also become the focus of blame. Recourse to genetics can express a sense of fatalism—"the luck of the draw"—or a moral judgment—there are "good" and "bad" genes.

Clearly, the gene of popular culture is not a biological entity. Though it *refers* to a biological construct and derives its cultural power from science, its symbolic meaning is indepen-dent of biological definitions. The gene is, rather, a symbol, a metaphor, a convenient way to define personhood, identity, and relationships in socially meaningful ways. The gene is used, of course, to explain health and disease. But it is also a way to talk about guilt and responsibility, power and privilege, intellectual or emotional status. It has become a supergene, used to judge the morality or rightness of social systems and to explore the forces that will shape the human future.

Our study examines both the form and the implications of the DNA mystique. We begin in Chapter 2 by exploring the social and biological powers of an earlier form of the gene, the so-called germplasm—heredity material that was invoked in the popular literature of the American eugenics movement

17

in the early twentieth century. Stories of the germplasm, so similar in its powers to the gene of the 1990s, suggest that ideas about heredity have as much to do with social meaning as they do with scientific research. In Chapter 3 we approach the problem of historical origins from a different perspective. Linking popular imagery of the gene to the medieval Christian conception of the immortal soul, we show how in many sources the gene appears to contain the essential human self and to promise a form of eternal life, either through children and grandchildren or through the body reconstituted or made "present" by scientific manipulation of DNA. Such historical references—to both eugenic and Christian ideas—help to locate the gene as a social icon as well as a biological entity. They frame our examination of the contemporary powers of the gene.

Chapter 4 examines the new "molecular family" as growing importance is placed on shared DNA. Chapter 5 shows that genes are common explanations for good and bad behavior or success and failure, and Chapter 6 explores how genetic images appear in the continuing public negotiation over issues of gender, race, and sexual orientation.

We then turn to the implications of genetic essentialism. Chapter 7 examines its ideological resonance in a society that tends to deflect responsibility for problems away from social or structural conditions and toward the individual. In this context, we argue, genetic essentialism becomes a theory of responsibility and blame. Moving to this theory's practical significance, Chapter 8 examines how routine acceptance of genetic assumptions influences institutional decisions, as employers, insurers, educators, and the courts apply genetic information to meet immediate economic and administrative needs. Finally, in Chapter 9, we use our data to shed light on questions that have preoccupied many observers of the current directions in genetic research. Does this science portend a "new eugenics"? To what extent is the cultural construction of genes and DNA conducive to eugenic social policy? We conclude in Chapter 10 by reviewing some reasons for the growing appeal of genetic essentialism and its significance in the context of both changing technologies for manipulating DNA and shifting social priorities.

2

The Eugenic Gene

In 1907 the "plant wizard" Luther Burbank wrote that "stored within heredity are all joys, sorrows, loves, hates, music, art, temples, palaces, pyramids, hovels, kings, queens, paupers, bards, prophets and philosophers . . . and all the mysteries of the universe."[1] In his popular 1934 eugenics primer *Like Begets Like*, Harry H. Cook told his readers that "within the nucleus of the germ cell lie the most important things in the whole world, the chromosomes, which are the determiners of character and in reality responsible for our natural individuality."[2] And in their popular 1920 text, California eugenicists Paul Popenoe and Roswell Hill Johnson announced that "immortality" was a "real possibility," for the germplasm, carrying the "very soul" of the individual, lived on in children and grandchildren.[3]

Appearing in texts written for a popular audience during the heyday of the American eugenics movement, these images of the germplasm (the term then used for the hereditary material) are strikingly parallel to images of the gene in the popular literature of the 1990s.[4] The germplasm, like the gene, was the determiner of character and personality, the source of social order, and the locus of immortality. Like the gene, the germplasm was not only a scientific concept but also a cultural resource, invested with spiritual and moral meaning. The germplasm was used to explore social problems and guide solutions to them. Among the woes attributed to the germplasm were criminality, mental illness, alcoholism, and poverty. Among the solutions favored by popular writers was the regulation of human reproduction.

The popular literature on eugenics from 1900 to 1935 was vast.[5] Our informal survey of this period has turned up at least 500 American titles written by nonscientists and apparently intended for a general audience.[6] Yet our numbers may be far too low. A 1924 bibliography of eugenics literature listed more than 4000 publications, approximately 1600 of which were popular texts or articles published in the United States between 1890 and 1924. The editor complained that much of this material was "uncritical; and comparatively little of it is written with the competence of the highly trained specialist."[7] It is precisely this "uncritical" material, however, that provides a way to understand popular constructions of the powers of the germplasm.[8]

Eugenics was not a single idea but a thousand ideas, not a simple, coherent doctrine but a messy public discussion that served many agendas. The volume and diversity of the texts suggest that eugenics interested many people and that it was taken seriously—by Midwestern housewives, Baptist ministers, labor unionists, elementary-school teachers, and pig breeders. Popular publications included Felicia Folger's privately printed *Great Mothers*, Max Reichler's *Jewish Eugenics*, physician Lawrence F. Flick's *Eugenics* (a lecture delivered in a Catholic summer school extension course), Scott Nearing's *The Super Race: An American Problem*, and the

20

writings of animal breeders such as Harvey Ernest Jordan (*Eugenics: The Rearing of the Human Thoroughbred*) and William Earl Dodge Stokes (*The Right to Be Well-Born, or, Horse-Breeding in its Relation to Eugenics*).[9]

Several famous figures promoted eugenics: plant breeder Luther Burbank, inventor Alexander Graham Bell, Stanford University president David Starr Jordan, and industrialist John Kellogg, among others.[10] And a 1928 survey of 499 colleges and universities found that 343 offered courses in genetics and eugenics.[11] Any attempt to assess the American eugenics movement historically must take this broad popular interest into consideration,[12] for amateur productions shed light on the meaning of the gene in the early part of the century, just as contemporary television programming and advertising campaigns reveal the symbolic meaning of the gene today.

The similarity between the powers ascribed to the germplasm in this early literature and the powers ascribed to the contemporary gene is intriguing. But the modern gene is not simply a revised and updated germplasm. Their shared powers rather reflect enduring Western ideas about self, social order, bodily difference, and family relationships. Both the germplasm and the gene have been used to address questions about the origins of social problems, the essence of the human condition, and "the nature of the very soul." By exploring the popular eugenics literature, we document the powers of the germplasm, closing with an examination of the persistence of biological ideas in American popular culture in the twentieth century. We suggest that the powers attributed to heredity in both the historical and contemporary contexts reflect cultural and social agendas more than they do the state of scientific knowledge.

Criminal Bodies

In G. Frank Lydston's 1912 play *The Blood of the Fathers*, a socially aware young doctor, interested in the hereditary theories proposed by nineteenth-century Italian physician and criminologist Cesare Lombroso, discovers that his own romantic choice is dysgenic.[13] The doctor rejects the love of an idealistic, high-minded social worker of impeccable breeding, falling in love instead with a charming patient, supposedly the child of an Army officer killed in the Philippines. Just before their marriage, the doctor finds out that his bride-to-be is actually the biological daughter of a convicted murderer and an opium-addicted mother who later committed suicide. Confronted with the "horrible outlook which only his knowledge of degeneracy and of hereditary criminality can give," he undergoes "an awful conflict of emotions." But "the heart wins; the man triumphs over the scientist, and the doctor marries" the woman.

After their marriage his bride, while attending a fashionable reception and ball, notices a valuable ornament in the hair of an elderly woman. "Her blood tells and she instinctively filches the jewel." Confronted with her crime, she "goes to pieces," throws the jewel on the floor, and then "yields to the sudden impulse to destroy herself" like her mother. "Alone with his sorrow" and aware that he "has made a mess of his life," the doctor vows that in his next marriage (perhaps to the idealistic social worker?) he will use "his brains" instead of his heart, knowing that "where other conditions are right, love for that mate is inevitable, and with the firm conviction that love never can survive mismating."[14] In the final scene summarizing the moral of the story ("The blood of the fathers!"), he addresses his dead wife: "You were wiser than you knew. You set things right—and you did it in the only way.—The blood of the fathers!—And our children yet unborn—and our children's children—they too, thank God! are saved—and in the only way."[15]

The "born criminal" appeared in many forums in the popular eugenics literature. In a 1912 booklet, feminist La Reine Helen Baker wrote that the "hereditary nature of the taint of criminality is proved by the history and bodily characteristics of its unhappy victims"; crime was not "a mere chance occurrence," nor was "badness the deliberate choice of the willful." People were rather driven by their heredity to commit crimes, helpless to stop themselves.[16] Similarly, in

OFFENSE RANKINGS OF RACIAL TYPES
Burglary, Larceny, Fraud

1917 science writer T. W. Shannon proclaimed that the "hereditary nature of the criminal propensity is unquestionable."[17] And Chicago medical school professor William S. Sadler observed in 1922 that "modern science is making clear to us that a large part of the criminality and immorality of the world, together with much of the alcoholic excesses, are due to [the] inherited taint of feeblemindedness." Sadler cited figures documenting increases in insanity that "seem to show conclusively that in just 330 years every man, woman and child in this country will be a raving maniac."[18]

The stories of inherited criminality and insanity presented in popular narratives suggested the legitimacy of state intervention, given the public costs of maintaining criminals in prison or caring for the "feebleminded." A chart used at the 1920 Kansas Free Fair presented a list of "unfit human traits" that included "feeblemindedness, epilepsy, criminality, insanity, alcoholism" and "pauperism." If all marriages were eugenic, the chart claimed, "we could breed out most of this unfitness in three generations." A placard for another exhibit asked: "How long are we Americans to be so careful for the pedigree of our pigs and chickens and cattle—and then leave the *ancestry of our children* to chance or blind sentiment?"[19] Equating reproductive control with crime control, the campaign to garner public support for eugenics turned biological improvement into a critical national need.[20] "The fate of the nation" depended on chromosomes,[21] and the "master key of history" was heredity.[22]

Pointing out that the first group officially recognized in the United States as a dysgenic threat consisted of poor women who were fertile, historian Nicole Hahn Rafter has characterized the early eugenics program as an attempt to "criminalize not an action but the body itself." Rafter's analysis of the establishment of the Newark (New York) Custodial Asylum for Feeble-minded Women in 1878 explores how fertile, poor women came to be seen as a social problem. She suggests that the image guiding "prophylactic" incarceration of poor women in the 1870s focused on the female body as a site of production. A promiscuous female who violated social norms was by definition feeble-minded. Her sexual behavior proved her to be flawed in body and mind, and women thus identified were institutionalized not to be cared for, but to be prevented from further reproduction; at menopause, they were commonly freed.[23]

The idea that the state had a direct stake in the control of the female body was reflected in the passage of state sterilization laws in the 1910s and again in the famous 1927 Supreme Court case, *Buck v. Bell*. In that case, Supreme Court Justice Oliver Wendell Holmes approved the sterilization of a teenage mother and her young child on the grounds that "three generations of imbeciles are enough." The case was orchestrated by eugenic legal reformers who wanted a high-profile test of the judicial strength of Virginia's new sterilization law. Their wishes were fulfilled, and the resulting courtroom drama drew on notions of a criminalized female body—a site for the production of defective goods— that were already well-entrenched in American legal circles.[24]

Eugenic anxieties extended from the female body to the family. The large dysgenic families described in the popular literature provided evidence for the "immutable law of God that like produces like."[25] Some of the family narratives, Rafter has observed, became "enormously popular, reaching wide audiences; others could have had but few readers. Cumulatively, however, they created a powerful myth about the somatic nature of social problems."[26]

Pathological Families

The first of such family studies in the United States was of the Jukes family, containing 18 brothelkeepers, 128 prostitutes, 76 convicted criminals, and 200 recipients of public relief. Richard Dugdale, the New York City merchant who discovered the Jukeses in 1874, estimated that this single family had cost the public more than $1.3 million (in public support and lost productivity). Dugdale himself recommended environmental improvement, "for where the environment changes in youth, the characteristics of heredity may be measureably altered."[27]

In 1897, eugenicist Elizabeth Kite began working with another family, named in publications "the Kallikaks," then living in the pine barrens of New Jersey. This family history included 143 feeble-minded persons, 26 illegitimate children, 33 prostitutes, 24 alcoholics, 3 criminals, and 82 infant deaths. In 1912, the director of the Vineland Training School, Henry Goddard, published *The Kallikak Family*, a popular text that presented this story of family degeneracy as self-evident proof of the inheritance of mental and moral inadequacy.[28]

In an orgy of data collection, eugenics field workers—most of them women—compiled family pedigrees stretching back two and three centuries and including hundreds and sometimes thousands of individuals. It was as though the quantity of data could demonstrate the validity of the assumptions that guided its collection.[29] Combining genealogical research with interviews and field visits, the workers reported on large family groups, all assigned pseudonyms, in which specific debilities seemed to predominate: alcoholism, crime, laziness, poverty, feeblemindedness, harlotry, locquacity, and so on. Some entire families lacked "emotional control" or "impulse control," others lacked industriousness (and were therefore doomed to be poor). The Jukeses were, by and large, criminals; the Ishmaelites, hereditary paupers; the Zeros, vagrants; and the Nams, alcoholics. The Hill Folk were "immoral" (particularly the women). All of the Dacks were emotionally unstable; all of the Kallikaks, feeble-minded. "The aberrant behavior of each family group is stamped with its peculiar characteristics," wrote Charles Davenport in his preface to *The Dack Family*. "In the Dacks we have a group of hyperkinetics whose reactions to their environment—a harsh environment to be sure which their reactions have tended to make such—are restlessness, quarrelsomeness, loquacity, abuse, pugnacity, intermittent outbursts of violent temper and sex offense." He predicted that the status of the family would go "from bad to worse until the natural process of purification by high infant mortality and by sterility do their beneficent work"—or unless the state intervened.[30]

The authors of the family studies tacitly acknowledged that the degenerate status of the families could be the product of environment—indeed, the rural environment of most of the families studied appeared in these texts to be just as pathological, in its way, as the family members themselves. But many of the families included individuals who had migrated to other locations, and the eugenics writers felt it was important to show that even when family members moved to more amenable situations, they continued to exhibit their familial taint. The family accounts "abundantly demonstrated that much, if not most, of their trouble is the outcome of bad heredity," said one popular eugenics text, for transplanted members of such families would soon "create for [themselves] an environment similar" to their original home. The "debasing atmosphere[s]" in which they lived, eugenicists argued, was a consequence of their inherent biological inadequacy.[31]

These texts presented complex human characteristics as simple inherited traits, either clearly dominant or clearly recessive, expressed in families in proper quantitative proportions. Thus in the Nam family of upstate New York, "lazy" parents produced "lazy" children in ratios demonstrating that laziness was a dominant trait.[32] Some members of the family were also the victims of "hereditary" alcoholism, prostitution, and shyness. But in one respect, the Nams were biologically quite healthy: They were significantly more fecund than the "most cultured families of the Eastern United States."[33]

The family studies were thus also an indictment of biologically superior families, believed by many eugenicists to be failing in their social responsibility by having so few children. Eugenics promoters wished to encourage such families to reproduce, and this encouragement took the form of public contests in which the powers of the germplasm to produce useful citizens were precisely quantified to help impartial judges assess "good stock."

Better Babies

During the First National Conference on Race Betterment, held in January 1914 at Battle Creek, Michigan, a series of "mental and physical perfection contests" attracted national publicity. These contests identified "better babies" and "perfect school children," based on physical measurements and mental tests. The winning babies received certificates or medals and public accolades. According to the conference report, "One interesting outcome of the baby contest was the adoption, because of the good record it made, of a baby that had been abandoned by its father."[34]

The awards at Battle Creek were part of a nationwide system of contests sponsored by the American Eugenics Society to promote its cause. Such contests called attention to eugenically desirable characteristics. The contest judges selected the physical and mental qualities that mattered, helping participants see themselves as competitors in a biological drama. The "fitter families" celebrated at county fairs, the "best baby" prizes awarded at the Race Betterment Conference, and the silver and gold medals given to young single people who were about to embark on marriage helped to demonstrate the eugenic ideal to the general public and to suggest what sort of stock was likely to produce eugenic children.[35]

Purebred farm animals were an obvious and widely used analogy for better babies, and the prize ribbons and medals

were like the standard awards of the stock show. As in judging cattle, evaluating babies involved comparisons of individuals against an apparently objective set of measurable qualities. What were these qualities? For infants and toddlers, the most important characteristic was developmental precocity. Thus six-month-old Virginia June Nay won in the infant category of the Better Babies Contest of 1914 because she was already teething and her fontanel had closed quickly. Her appearance, which was pleasant, was not a factor in her success. "You will note that the most perfect baby and not the most beautiful baby is selected. The possession or lack of beauty does not score either for or against a child," said the prize committee in its explanation of the awards. Disposition, too, was irrelevant. Cranky babies could win prizes, for all that mattered were the objective measurements. Points were deducted for "any deviation from the norm" in development, size, or shape. "Every baby has been examined head to foot. A score card indicating the different measurements and examinations to be made has been followed in each case. By referring to your score card, you will note that a certain percent is deducted for each deviation from the normal, and that the child's final score is the result of the sum of these taken from one hundred percent."[36]

The Battle Creek meeting also included a contest for school children, who were ranked according to their scores on standardized tests, mental outlook, height, and various physical traits such as dental development, vision, and hearing acuity. The "Mental and Physical Perfection Contests" were a testament to the persuasive power of numbers, for like the better babies contests, they were decided entirely on quantitative scores. Thus a boy with "normal lung capacity for age twelve," calculated as 116 cubic inches, received a score of 12 points; one with 128 cubic inches received 13 points. Desirable height was based on the "Hastings' Age-Height Tables." Children who deviated from these tables in either direction were eliminated from the competition. The contest judges had also planned to eliminate those without perfect teeth, but only sixteen of the 3,537 school children examined could survive such an elimination, so the criterion was dropped.[37]

The report of the 1914 Race Betterment Conference noted the "remarkable interest" of the press in their contests. Reporters arrived representing the Associated Press, the United Press, the Scripps-McRae League, the Detroit *Free Press*, the Chicago *Tribune*, the Detroit *News Tribune*, the New York *World* and *Herald*, and the Philadelphia *Record*. In addition, the Pathé Weekly moving picture firm sent a crew to cover the meetings, and cables brought news of the event to newspapers in London and Paris. After the conference, reports appeared in "medical, scientific, economic and other journals throughout the country."[38]

Perhaps inspired by this enthusiasm, the American Eugenics Society later sponsored similar contests at state fairs, in which "fitter families" were identified and rewarded for their physical and emotional superiority. The prizes made explicit the idea that desirable hereditary traits could be deciphered simply by an examination of the phenotype.[39] Any family interested in competing had to agree that all of its members could be examined by a physician and that all would take an intelligence test. The "Fitter Families Examination" form asked the examiner to grade both individual family members and the entire family on a scale of A to C ("use + or − if necessary") for such traits as "good provider," "generous," "self-sacrificing," and "walks or talks in sleep." Examiners were to use "well-sharpened red, blue, and black pencils." Blue was "to indicate excellence or superiority," red to mark "all defects and penalties," and black for other entries.[40]

Fitter families were the logical counterpoint to the Jukeses and the Kallikacks, and their consistent goodness demonstrated the powers of the germplasm with just as much force as did the consistent weakness of the degenerate families. The central character of the eugenics movement was the "germ-plasm," the essential entity through which a social transformation would be achieved. And the agents of this social transformation would be the individuals who were "well-born," the "better babies" celebrated in the public contests and competitions sponsored by the American Eugenics Society in the 1920s.

The stories of unsavory families marked by tragedy and social dissolution, the advice to parents and parents-to-be, and the fitter families contests were often intended to persuade an educated but not necessarily scientific audience of the need for progressive social reform. American promoters of eugenics sought to convince the public that the controlled breeding of human beings was morally acceptable, socially progressive, and scientifically legitimate.

A Civic Religion

Eugenics in the United States was not merely an elite debate about population genetics or statistical methods; nor was it simply institutionalized nativism or racism. Eugenics was a civic religion, and its amateur promoters did not need data provided by the biometricians or the Mendelians, the Darwinians or the neo-Lamarckians to support their beliefs. They needed only the well-bred milk cow, and the increasingly sacralized child.[41] Just as no cattle breeder would permit his worst bull to sire the next generation, so no society should permit its weakest citizens to become parents. This was the model for scientific management of the human gene pool as it was represented in the popular eugenics literature.

Yet the prospect of efficient population management alone was insufficient to sell the movement's agenda to the broader public. Some were troubled by the apparent contradictions between Christian morality and the aggressive social control of reproduction. Given the prevailing prejudice against Darwinism, played out spectacularly in the 1925 Scopes trial, it is not surprising that Christian writers believed that their readers would be resistant to eugenics on moral grounds.[42] Some tried to assure their readers that eugenics was not immoral but a matter of "rights." A 1914 American Baptist Publication Society text on the "rights of the unborn race" insisted on a baby's right to a sound body, not his or her rights to education, nutrition, or medical care.[43] One of the questions listed in the 1926 *Eugenics Primer* asked: "Does eugenics mean less sympathy for the unfortunate?" The answer: "Eugenics does not mean less sympathy for the unfortunate; it does mean fewer unavoidable unfortunates with which to divide a sympathy. . . . This is true kindness, both to the victims and to society."[44]

Many eugenicists were emotionally committed, drawn to the movement by a sense of social mission and a vision of a human future free of disease, mental illness, and criminal inclination—a vision of healthy babies and of the right of the child to be "well-born." Many popular texts therefore cast eugenics in utopian and dystopian terms. Action now, they said, would produce a race of Michaelangelos and Shakespeares; inaction would lead to human extinction. Many promoters were attracted to the vague scientism of eugenics, but they were also drawn to its potential for transforming the human condition and for eliminating the everyday human suffering caused by physical or mental disability.

Eugenicists also appealed to economic concerns about the costs of caring for the degenerate. They assured their readers that control of who was allowed to reproduce—controlling the qualities of the collective germplasm—would help reduce public costs. Eugenics could, they promised, reduce the number of unproductive members of society and increase the number of intelligent and ethical individuals, thereby improving, as one popular lecture termed it, "the world's most valuable crop."[45] Such narratives, conforming to the social agendas of the era, could easily be appropriated by diverse groups to meet their purposes and solve their problems. The malleability of eugenic narratives contributed to their public profile.

Strategic public support for eugenics did not necessarily require an understanding of, or even an interest in, heredity: When eugenics narratives conformed to social interests and public concerns, they were readily appropriated. Thus eugenic ideas became part of the arsenal of public officials concerned about the economic burden of supporting the poor, labor interests threatened by cheap immigrant workers, and an elite group interested in maintaining class privilege. In many ways, concerns about the details of heredity—such as the risks of species decline or the dangers of long-term human devolution—were secondary to the social agendas that eugenics addressed.[46]

At the same time, some who accepted an extreme hereditarianism were suspicious of the practice of eugenics, which appeared to be administered by a politically corrupt, socially imperfect system. The problems of eugenic policies were the subject of the 1934 film *Tomorrow's Children*, a story about the efforts of a state welfare worker to have the healthy eldest daughter of a dysgenic family sterilized. The young woman is hardworking, attractive, and kind, but her parents are alcoholics and her younger siblings suffer from mental or physical handicaps. The film presents virtually all the flaws of the family members as hereditary. It does not question the basic premise of eugenics that such traits as "laziness" or "tendency to steal" were inherited. But it doubts the fairness of the social systems through which eugenics would be enacted. It also dramatizes, in biological terms, the problem of the deserving poor.

In a courtroom scene, a malevolent-looking young man with a clear history of violent acts is spared the surgeon's knife when his influential attorney, who is on a first-name basis with the judge, reveals that the man's father is a powerful local politician. Meanwhile the hardworking young woman is ordered to the hospital for sterilization, then saved at the last possible moment by the mother's drunken confession that her daughter had been adopted as an infant. The film suggested that in practice those with power would be spared regardless of their biological fitness, with tragic consequences. *Tomorrow's Children* is both a rabidly hereditarian and an anti-eugenic film.[47]

The Shifting Fortunes of Eugenics

The idea that human beings could and should be selectively bred was emotionally fueled by the fear that differential reproduction of the poor and unfit would doom the species to biological decline, coupled with the dream that a program of controlled breeding could produce a super-race of Michaelangelos and Newtons. These tightly linked ideas of biological decline and biological progress reached their ultimate expression in the eugenic policies of Nazi Germany, where the notion of genetic purity in the Aryan race became the justification for the racial hygiene movement. Fit individuals were encouraged to reproduce, and those judged unfit were sterilized or murdered.[48]

The eugenics movement declined in the United States after 1935, its social agenda called into question by the unfolding events in Germany and its scientific legitimacy challenged by population geneticists, anthropologists, and behavioral psychologists.[49] After World War II, the revelations of Nazi atrocities and the liberation of the death camps helped to encourage a general cultural shift in which environment seemed to explain human behavior and human difference more completely than did heredity.

Nurture triumphed indisputably in both the scientific and popular rhetoric of the 1950s and much of the 1960s. The stories of biological determinism that had characterized the eugenics literature were replaced by narratives of cultural determinism. Adults appeared in the popular press as products solely of their individual experiences. Indeed even the newborn, in the popular literature of the 1950s, was already a cultural product, shaped by experience in the womb.[50] As late as 1972, a social scientist could write that: "The idea that man has no nature is now beyond dispute. He has, or rather is, a history."[51]

But among some professional groups the vision that fueled the eugenics movement persisted throughout the 1950s and 1960s. During this period, expectations of the medical and social benefits of regulating reproduction had a prominent (and little-explored) place in both infertility research and scientific human genetics. The persistence of eugenic language in these two professional arenas is important to us because of their later significance in the development of new technologies (such as genetic diagnostics) linked to the control of reproduction.

In 1950 the American Society for the Study of Sterility began publishing a new journal, *Fertility and Sterility,* with two stated goals. The first was to "assist clinicians in the treatment of infertile couples." The second was to "discover not merely how more, but how finer people can be bred. To improve the quality of man . . . should be the ultimate goal of all investigations of fertility and infertility."[52] The journal's clientele apparently embraced this goal. In a 1954 essay on the implications of donor insemination, fertility researcher Alan F. Guttmacher noted that scientists could help replace "tainted gametes transmitting cacogenic [negative] qualities [with] those transmitting eugenic qualities" by carefully choosing donors.[53]

Eugenic ideas also continued to shape the professional development of human genetics. In the fall of 1961, twelve geneticists met at Princeton for the Third Macy Conference to discuss the workings—or rather, the inefficiency—of natural selection in human populations.[54] Like eugenicists early in the century, these leading scientists were concerned about the apparent suspension of natural selection made possible by improvements in medical care and the standard of living. What would it mean for the future of the species? The discussion of this issue in the professional literature of the 1960s was not about the elimination of human suffering; those attending the Macy Conference postulated, rather, that inbreeding has a "purifying" effect on human populations, since it permits deleterious recessives to be "exposed" and "outed" by the birth of defective children who are unable to reproduce.[55] The concerns that prompted such discussions had more to do with the social and medical costs of maintaining persons born with genetic disease than with their medical needs—more to do with the future health of the human gene pool than the present health of patients.

Victor McKusick, one of the leading pioneers in scientific human genetics, noted in his 1964 textbook that eugenics had a "laudable objective" but had "brought ill repute to the science of genetics." The "perversion of eugenics in the racist philosophy of the Nationalist Socialist regime of Germany undoubtedly caused a significant setback." Like many others, McKusick favored eugenics generally for its potential to reduce the incidence of serious genetic disease, but objected that not enough was known to make a eugenics program practical. "There is not enough scientific information on which to base recommendations for large-scale eugenic action."[56] And in 1974 Frederick Osborne, president of the Society for the Study of Social Biology (the new name, circa 1972, of the American Eugenics Society), said that the drive for a "valid eugenics" was going well, that smarter people were having more children, and that genetic counseling and "wanted" babies were making possible more eugenic choices. He noted that the society was sponsoring scientific conferences on the course of human evolution, differential reproduction, and the evolution of races, and that members hoped that such meetings would help scientists understand how to influence "the future course of human evolution."[57] The notion that more scientific information about human heredity would make possible a valid eugenics continues to appear in the writings of prominent scientists, even in the 1990s (see Chapter 10).

The post–World War II story of biological determinism shared with the earlier eugenics movement a fascination with the bodily origins of social problems. The image of a perfected population—the body politic purified through self-directed evolution—continued to exert a powerful cultural appeal.

The urgent reasons to control the germplasm in the early eugenics literature had focused on social cost, rather than on private suffering. The hereditary material in these accounts was not yet medicalized—individual disease per se was rarely the specific issue. But the nonmedicalized and nonspecific "germplasm" that played such an important role in the American eugenics movements had many of the same powers and characteristics as the contemporary gene. The germplasm was a source of identity, the determiner of fate. There were good stocks and bad stocks, hereditary intelligence and feeblemindedness, special talents and criminality. Indeed, the almost magical powers of the germplasm resonate in remarkable ways with those of the highly medicalized and specific gene of the 1990s.

Images from the early eugenics literature have reappeared, though often in different guise. "Better babies" are still a highly desired reproductive commodity, if not the subject of nationally publicized contests. "Born criminals" are targeted by public health initiatives that focus on the "urban underclass." The dysgenic woman incarcerated at the Newark Custodial Asylum for Feeble-minded Women has become the "worthless welfare mother with numerous children destined to join the ranks of the unemployed and criminal."[58] And the problems of American competitiveness and economic stability seem, in some texts, to turn on the differential reproduction of classes and races.[59] While the contexts are different, the emphasis on the body and its role in generating social problems persists.

For most of the twentieth century, with the exception of the two decades following the 1939–1945 war, biological and genetic explanations of human behavior have captured the American popular imagination. But in the 1990s they have attained a new legitimacy and currency that draw on the newest advances in molecular genetics, the primal concerns with the body, and the fears about the human future that had found early expression in the popular narratives of the American eugenics movement.

"Life everlasting," wrote influential California eugenicists Paul Popenoe and Roswell Hill Johnson in 1920, "is something more than a figure of speech or a theological concept" for the eugenicist. "The death of the huge agglomeration of highly specialized body cells is a matter of little consequence" if the germ-plasm has been passed on, because the germplasm contained the "very soul" of the individual.[60] In the 1990s, the "very soul" placed in the germplasm by Popenoe and Johnson has moved to the DNA, as popular treatments of personhood increasingly draw on genetic images to explore questions of immortality, identity, and fate.

3

Sacred
DNA

In Stephen Spielberg's popular 1982 film *E.T.*, the extra-terrestrial hero, apparently dying, lies on an operating table; suddenly a scientist runs in shouting, "He's got DNA!" Like many other cues in the widely admired movie, this reference to E.T.'s DNA reflects familiar ideas. It is part of a cultural narrative in which DNA is removed from history: this essential molecule is seen, not as a consequence of the conditions under which life evolved on Earth, but as an entity present in all living things regardless of their planet of origin. Indeed, discovering DNA in E.T.'s body is analogous to finding the King James Bible in the hold of a Martian spaceship. Such a discovery liberates the molecular text from history and makes it seem truly universal.

The scientific world view is based on belief in an underlying order in nature, and many scientists search, with nearly religious conviction, for an ultimate, unifying principle that will reveal the most fundamental laws.[1] Physicists in particular have interpreted their work in cosmic terms. Stephen Hawking, in *A Brief History of Time*, proclaimed that scientists reveal "the mind of God."[2] Nobelist Steven Weinberg, in *Dreams of a Final Theory*, searched for the principles that would explain all the laws of nature.[3] Physicist George Smoot has compared the big-bang theory to "the driving mechanism for the universe, and isn't that what God is?"[4] And Leon Lederman, another Nobel Prize-winning physicist, has named the subatomic entity that he believes determines everything the "God Particle." He has stated that he hopes to see all of physics reduced to a formula so simple and so elegant it will fit on a T-shirt.[5]

Biologists, too, have sought to unify biological knowledge through elucidation of the fundamental properties of life. In the 1930s in Britain and the United States, this effort took the form of the "evolutionary synthesis," which seemed to reconcile Darwinism and Mendelism—selectionism and genetics—theories initially seen as contradictory. The architects of the synthesis were able to promote the idea that biological change through time—evolution—could serve as the intellectual centerpiece for the study of life.[6] In the same period, the rise of molecular biology promised to explain life at its most fundamental physico-chemical level, the double helix of DNA.[7] And in 1975 entomologist E. O. Wilson announced a "new synthesis" that drew on both evolutionary biology and molecular biology to explain the human social order in biological terms.[8]

One of the most important entities in the search for an essential, unifying biological principle, then, has been DNA, the so-called "secret of life." In the 1990s geneticists, describing the genome as the "Bible," the "Book of Man," and the "Holy Grail," convey an image of this molecular structure not only as a powerful biological entity but also as a sacred text that can explain the natural and moral order. Former director of the Human Genome Project and Nobelist

James Watson has proclaimed that DNA is "what makes us human."[9] "Is DNA God?" asks a skeptical medical student in an essay in *The Pharos,* a medical journal: "Given [its] essential roles in the origin, evolution and maintenance of life, it is tempting to wonder if this twisted sugar string of purine and pyrimidine base beads is, in fact, God."[10]

Such spiritual imagery sets the tone for popular accounts of DNA, fueling narratives of genetic essentialism and giving mystical powers to a molecular structure. Indeed, DNA has assumed a cultural meaning similar to that of the Biblical soul. It has become a sacred entity, a way to explore fundamental questions about human life, to define the essence of human existence, and to imagine immortality. Like the Christian soul, DNA is an invisible but material entity, an "extract of the body" that has "permanence leading to immortality."[11] And like the Christian soul, DNA seems relevant to concerns about morality, personhood, and social place.

It is not a coincidence that the cultural depiction of DNA shares many characteristics with the immortal soul of Christian thought; those describing DNA often draw on the most powerful images of Christianity to convey its importance. Scientists and popularizers borrow the compelling concepts of one belief system to meet the needs of another in an effort to help their readers see the centrality and power of the gene.

Most cultures have recognized some entity that is relatively independent of the body, but that gives the body life and power.[12] Known in various historical and cultural circumstances as the soul, *yalo, nöos, hun,* spirit, and so on, this entity persists when the body is gone and, containing all its essential elements, can be used to bring the body back (for example on the day of the resurrection of the dead, the final day of judgment). This independent entity is also central to identity or selfhood; as philosopher Richard Swinburne has observed in his study of the nature of the soul, "personal identity is constituted by sameness of soul."[13]

So, too, in contemporary American popular culture, DNA is relatively independent of the body, gives the body life and power, and is the point at which true identity (and self) can be determined. DNA, like the soul, bears the marks of good and evil: A man may look fine to the outside world, but despite appearances, if he is evil, it will be marked in his soul—or his genes. And DNA also appears to be immortal, containing within it everything needed to bring the body back. Cloning DNA has become, in popular culture, the way to reconstruct the bodies of extinct organisms (the dinosaurs in Michael Crichton's *Jurassic Park*), or to resurrect the characteristics of such past heroes as Abraham Lincoln.

Stories of DNA in popular culture also incorporate the classic myths of Frankenstein or the garden of Eden. Modern molecular genetics promises a "complete" understanding of human life, but such promised knowledge, in the form of genetic engineering and genetic therapy, also commonly appears as dangerous and taboo. Manipulating DNA, in this view, becomes a sacrilege, a violation of sacred ground.

In his analysis of the "sacred and profane," anthropologist Mircea Eliade describes how sacred realities become manifest "in objects that are an integral part of our natural 'profane' world." Thus, human organs in many cultures have been sacralized—endowed with religious valorization.[14] And in a study of theological debates about the soul in the twelfth and thirteenth centuries, historian Carolyn Bynum shows how the issue of personal continuity—the survival of the self or soul—has long focused on actual physical body parts. Questioning "how identity lasts through corruption and reassemblage" of the body, early Christian thinkers debated whether discarded fingernail parings would be reunited with their rightful owners at the end of the world.[15]

The modern cultural concept of genetic essentialism draws much of its power from such theological roots. The gene has become a way to talk about the boundaries of personhood, the nature of immortality, and the sacred meaning of life in ways that parallel theological narratives. Just as the Christian soul has provided an archetypal concept through which to understand the person and the continuity of self, so DNA appears in popular culture as a soul-like entity, a holy and immortal relic, a forbidden territory. The similarity between the powers of DNA and those of the Christian soul, we suggest, is more than linguistic or metaphorical. DNA has taken on the social and cultural functions of the soul. It is the essential entity—the location of the true self—in the narratives of biological determinism.

Demarcating Boundaries

Anti-abortionists describe the base pairs of DNA as the "letters of a divine alphabet spell[ing] out the unique characteristics of a new individual" at the moment of conception.[16] For right-to-lifers, a complete set of chromosomes is a complete person. The chromosomes define and contain the individual in a "master genetic code." Just as genomics scientists characterize DNA as the "stuff of life," so religious leaders characterize it as a "core of essential humanity." In different ways, these groups are exploring the problem of boundaries: What is the crucial characteristic of humanity? What "makes us human"? That so many voices in the contemporary discourse on the essence of human life should settle on humanity's DNA as an answer to this age-old question is compelling evidence of the iconic importance of the gene as a secular equivalent to the soul. This concept provides biological grounding for the shifting and unsettling boundaries of identity in our time.

Societies commonly draw boundaries to define personal identity and human exceptionalism, but in the late twentieth century traditional demarcations are besieged. Theories of artificial intelligence suggest that human intelligence is not unique but can be experienced by "thinking machines."[17] Virtual reality devices fuse the biological with the mechanical, reducing human "experience" to stimulation of the neocortex.[18] Animal rights activists argue that humans are not exceptional, and therefore that the rights we enjoy should be extended to all other animals.[19] And the evolutionary narratives of sociobiology claim that human social behaviors and cognitive characteristics are simply an extension of those in primates.[20]

New words have entered the language to express the tension over such boundaries. "Cyberspace" is where the mind and computer chip embrace. "Cyberpunk bodies" are "spare, lean and temporary bodies whose social functionality [can] only be maintained through reconstructive enhancements—boosterware, biochip wetware, cyberoptics, bioplastic circuitry, designer drugs, nerve amplifiers, prosthetic limbs and organs, memoryware, neural interface plugs and the like. The body [is] a switching system, with no purely organic identity."[21] The "cyborg"—a word coined in the 1960s to describe a cybernetic organism—stands, in Donna Haraway's formulation, at the "blurred and anxiety-inducing boundaries between human and animal and between organism and machine."[22]

Meanwhile scientific promoters of "biomimicry" predict "a car that could heal itself after a fender-bender" and aircraft exteriors that will be structured like rhino horn.[23] The cover of Bryan Appleyard's study of "science and the soul of modern man" features a robotic hand reaching out to touch a human hand, as in Michelangelo's depiction of the moment of creation.[24] And a computer program, SimLife, the Genetic Playground, promises students a chance to "design an ecosystem, populate it with imaginary plants and animals and, by introducing mutagens, cause havoc as species become extinct."[25] Such diverse images suggest that, in the 1990s, the lines once assumed to be clear cut between "man" and "nature," or "life" and "technology," have been shaded over and obscured, often in troubling and discomforting ways.

The traditional lines of class, race, and gender that once neatly divided the social world have become contentious in new ways, as well. As the old rules for dividing the world and defining one's place in it are undermined, genetic essentialism promises to resolve uncomfortable ambiguities and uncertainties. The genome appears as a "solid" and immutable structure that can mark the borders and police the boundaries between humans and animals, man and machine, self and other, "them" and "us."

The idea that those who have human DNA are irrepressibly human—regardless of how they have acquired that DNA—was implicit in Ridley Scott's 1982 film *Blade Runner,* re-released with a bleaker ending in 1992. Set in 2019 in post-apocalypse Los Angeles, the film explores the problematic status of the mutinous "replicants," short-term human clones created to perform the demeaning and dangerous jobs once performed by those assigned (because of their social class, race, or sex) to the lower strata of society. The replicants provide sex for hire, colonize dangerous territories, and fight wars. These clones supposedly have no real feelings—no fear and no shame—though they were programmed to express such emotions as desire when the expression would please their human users. Programmed to have a four-year life span, they are also believed to have no personal regrets or will to live. The plot of the movie, however, is built around their uncontrollable humanness; for the replicants—as clones with human DNA—want to live. They begin going AWOL in order to find the scientist responsible for creating them, with the aim of convincing him to help them live longer.

The story uses the replicants to explore the problem of ethnic and social class differences and the hopelessness of seeking technological solutions for social problems.[26] But it also explores questions about boundaries by presenting the replicants—manufactured humans concocted of manipulated DNA—as more "human" than the evil corporate planners who made them. Though constructed only to serve society's needs and exploited as slaves, the replicants have a fundamental will to survive. They have identity, selfhood, (false) memories of childhood, and hopes for the future. They are therefore fundamentally human.

Similarly, in a comic book series called *DNAgents*, the Matrix company creates synthetic human beings who look human and act human, but whose "DNA codes have been altered just enough to make them more than human, the perfect special agents to work for Matrix." The company sends the DNAgents on missions that do not always turn out as expected, for the agents prove to be independent, to have an irrepressible human essence. The message: human DNA demarcates the human from the robot, so even constructed

beings will claim human rights if they contain human DNA. As the comic book slogan notes, "Science has made them but no man owns DNAgents."

The same theme appears in the *X-Men* comic book series: shared DNA is the essential characteristic defining human-ness and justifying rights and respect. In the futuristic world of the X-Men, mutant humans with dorsal fins and tele-kinetic powers are the social equivalent of African Americans, Jews, Asians, and other minority groups. Their creator, Dr. Xavier, pleads with the public to accept the "muties," for "we are related, we are all family."[27]

In these science fiction narratives, shared DNA seems to permit the inclusion of those whose differences—in history or in bodily traits—mark them as outsiders. In other narratives, those who share the same DNA are dangerously close to being the same person. Anthropologist and popularizer of sociobiology Melvin Konner suggests that twins separated at birth, when reunited as adults, experience a "strange bound-ary-blurring union." Who am I? is one of the most basic human questions, Konner notes. Meeting another human being who is genetically identical is therefore, he says, a jarring experience, a challenge to self-actualization.[28] This idea was explored in the 1989 Irving Reitman science fiction comedy *Twins,* in which two brothers separated at birth discovered eerie similarities of habit and taste, despite their profound differences in body type and personal history. Created by scientists at Los Alamos, they were the result of an experimental effort to create a superman by using bits of sperm from six fathers chosen for their genetic excellence. But instead of a single superman the scientists got two baby boys, one endowed with all the desired traits, the other with the leftover "genetic garbage." The good "twin" was meti-culously raised on a tropical island, the other sent to an orphanage. But eventually they found each other and discov-ered they were the same; they had identical gestures and habits and could read each others' minds. While they looked very different, they shared essential qualities as a conse-quence of their status as "twins."

DNA as a boundary marker and a source of true identity has come to play a practical role through DNA fingerprint-ing. The public and judicial enthusiasm for this means of identification is further evidence that DNA has taken on a cultural meaning as the essence of the person, for popular descriptions of this scientific technique emphasize its awesome powers of sorting and identification.

In 1983 Alec Jeffries, a geneticist at the University of Leicester in England, proposed that the DNA contained in biological materials at a crime scene—dried blood or semen, for example—could be used to help identify who had committed the crime. The DNA could be cut up, separated by size, and then compared to a suspect's DNA (which had been similarly processed) for the presence of specific DNA sequences known to vary in human populations.[29] In 1987 Jeffries's technique helped solve a widely publicized British murder case, and since then so-called DNA fingerprinting has become a powerful form of evidence in the courts, used to document whether or not a given suspect was at the scene of the crime.

Press accounts called DNA fingerprinting the "single greatest forensic breakthrough since the advent of fingerprinting at the turn of the century," predicting it would "revolutionize the investigation of violent crimes." A spokesman for a biotechnology firm that conducted DNA fingerprinting announced that the possibility of error in identification was "one in 4 or 5 trillion" to one; a news magazine headline proclaimed DNA fingerprinting a "a foolproof crime test."[30]

In practice, however, DNA fingerprinting is a statistically reasonable—but not infallible—method of identification, and its use in court, as in the O. J. Simpson trial, has been contentious. Laboratories that produce DNA fingerprints are still struggling to control errors.[31] The commonly used DQ Alpha method of testing DNA is not particularly precise: the odds that two people will have the same combination of markers are currently estimated as ranging from one to 20 percent—quite different from the figure of 4 or 5 trillion to one commonly cited in the popular literature.[32] In popular stories, however, the DNA "fingerprint" appears as the "ultimate identifier," an utterly conclusive code establishing the essence as well as the identity of the person.

Such grandiose claims have made their way into various cultural arenas, appearing, for example, as the focus of a New York City gallery art exhibition in 1993. Conceptual artist Larry Miller offered gallery visitors a "Genetic Code

GENETIC CODE COPYRIGHT

I, _____,

Born a natural human being on the _____ day of _____ in the year _____,

In the town, state and country of_____

Of my mother, _____ and my father, _____,

DO HEREBY FOREVER COPYRIGHT MY UNIQUE GENETIC CODE,
HOWEVER IT MAY BE SCIENTIFICALLY DETERMINED, DESCRIBED OR OTHERWISE EMPIRICALLY EXPRESSED.

Any reproduction, regeneration or facsimile duplication, whether in whole or in part,
whether physically manifested or technologically represented is UNIVERSALLY PROHIBITED.
All rights and permissions are reserved and may only be assigned, in whole or in part,
to a specified agent via legal, written authorization by me or by my legal heirs upon my decease.

Sworn to and subscribed before me as witness
on this _____ day of _____
_____ in the year _____
in the town, state and country of

Sworn and declared by me,
an Original Human,
with fingerprint affixed herein

WITNESS

PROCLAIMER

GENETIC CODE CERTIFICATE © LARRY MILLER 1997

Copyright" and invited them to sign it: "I _____ being a natural born human being . . . do hereby forever copyright my unique genetic code, however it may be scientifically determined, described or otherwise empirically expressed," the certificate stated. "Sworn and declared by me, an original human, with fingerprint affixed herein." For $10, Miller would witness the completed forms for visitors.[33] Miller's copyright certificate satirically questioned the cultural construction of DNA as the immortal essence of "an original human." Camille Paglia, however, has taken this concept more seriously: "Behind the shifting faces of personality is a hard nugget of self, a genetic gift. . . . Biology is our hidden fate."[34]

In the plots of many popular stories, physical appearance is not enough to establish identity. A person may look just like, but not really *be*, the accused. Only their DNA, perceived as unique in each individual (except identical twins), can determine identity with certainty. For example, in a recent episode of "Deep Space Nine," the current prime-time spin-off of the "Star Trek" television series, an alien accused of murder is identified by his DNA, which was compared to a sample in a computer file similar to the FBI databank. Soap operas and made-for-TV movies, moreover, use DNA finger-printing in stories about efforts to establish identity— whether of a criminal, a suspected father, or a claimant for the family fortune.

In these narratives DNA becomes, in effect, a contempo-rary soul, the site of identity and self. The privileging of DNA tests and the cultural expectation that they can provide a virtually infallible way to identify individuals reflect the magical power attributed to DNA (and, by extension, to molecular genetics) in American popular culture. This power is even more explicit in the construction of DNA as a modern molecular relic.

The Genetic Relic

A contemporary molecular biologist—one of the pioneers of a technology widely used in genomics research—has founded a company that will produce cards or jewelry containing DNA cloned from musical superstars, athletes, and other secular saints. Kary Mullis, who won the 1993 Nobel Prize for developing the gene amplification technique called poly-merase chain reaction (PCR), explained to the *New York Times* that the purpose of such cards will be to educate people about DNA.[35] He has even proposed selling cards with DNA from various primates as a way to illustrate evolution for school children. "The idea is that teenagers might pay a little money to get a piece of jewelry, a bracelet or whatever, containing the actual piece of amplified DNA of somebody like a rock star," Mullis has said.[36] And along the way, they may learn a little molecular biology: "People could use the cards as totems or relics, but they could also learn about genes by comparing different stars' sequences."[37]

Mullis's DNA cards can be understood as a form of con-tagious magic, the mystical construct that, for example, underlay the widespread distribution of pieces of the True

Cross (on which Christ died) and other Christian relics in the fourth and fifth centuries. In contagious magic, any object that comes in contact with a revered person (or a part of that person's body, such as hair or bone) is believed to be equivalent to the person's whole self, no matter how small or how distant in time. A fragment of bone, a single hair, or a bit of cloth or wood from an object once touched by the person can, in the words of the *New Catholic Encyclopedia*, "carry the power or saintliness" of the person "and make him or her 'present' once again."[38] Such objects, commonly called relics, played an important role in early Christianity. At the height of the "cult of relics" fashionable noblewomen wore around their necks amulets containing such objects as a purported splinter of the True Cross. By the middle of the fourth century, wood from the True Cross "filled the world," though "miraculously the original cross remained whole and undiminished in Jerusalem."[39] The rage for relics had the advantage of bringing the saints directly to the people, and the remains of saints became a symbolic exchange commodity that fostered the spread of Christianity at a pivotal time in Church history.[40] They also became the basis of a brisk and lucrative trade in medieval relics, often enriching church officials.

Like the True Cross in the early Christian period, the bits of celebrity DNA produced by Kary Mullis and his company could "fill the world" without becoming depleted. "We just have to get a little piece of skin, clip a nail, or something from the person, prepare the DNA [and] copy it through PCR." The resulting bit of biological material could then be encased in bracelets, Mullis suggested. "You could say 'here is a sequence' from Mick Jagger, something to do with his lips, say. The jewelry will look like something your gypsy grandmother gave you and in there will be a little speck of DNA." A bit of DNA from a dead celebrity might be particularly appropriate, Mullis told *Omni* magazine. "If we could get permission to use someone like Elvis Presley, we could do a gene of the month, and you could have a collection like stamps." Instead of jewelry, however, the company decided to produce something similar to "a baseball card, with the person's picture and some of their DNA worked right into the card, and some sequence information printed on the back."

Mullis, like early Church leaders, is interested in spreading the faith by bringing celebrity DNA to the people. Molecular relics promise to make the revered person "present" for the follower. And, like relics in the fourth century, DNA cards will educate their owners, enrolling them in the molecular paradigm. Mullis is explicit about this agenda: comparing them to Christian relics, he intends the DNA cards to be a form of popular promotion of molecular genetics.

Molecular relics have also appeared in stories about the investigation of Lincoln's DNA. In February 1991, the National Museum of Health and Medicine appointed a committee to study the technical and ethical feasibility of obtaining DNA contained in bits of Lincoln's hair, bone, and blood stored in museums. Scholars have long theorized that Lincoln might have suffered from Marfan syndrome, a rare genetic condition characterized by weaknesses in the bones and joints, eyes, and heart. Anecdotal evidence links Marfan to high intelligence, and Marfan patients are often tall, with long limbs and fingers, fueling speculation that Lincoln suffered from this disease.

The primary risk in the condition is that the aorta will burst—many Marfan victims die relatively young as a consequence of heart problems. The historical debate about Lincoln as a victim of Marfan syndrome has explored whether the disease could have taken his life at any time even if John Wilkes Booth had failed to assassinate him in April of 1865. "Was the slain president doomed by a disease?" asked a headline in a *New York Times* account of the plan. The "genes define the essence of the person," noted one journalist covering the debate over Lincoln's DNA: "Some scientists suggest that genetic evidence might also one day show whether Lincoln suffered from chronic depression, as several biographers suspect, or from other conditions that affected his decision-making."[41]

In this narrative, President Abraham Lincoln—the entire social, historical, cultural, and biological actor—can be retrieved from relic-like body parts stored in museums in

Washington, D.C. His DNA seems to "make present" the historical figure in all his complexity. Molecular analysis of DNA can reveal the structure of his intelligence and his emotional state, even his decision-making style. And unlike Lincoln's own writings, his speeches, his correspondence or the correspondence of those who knew and observed him in action—unlike these archival documents chronicling his actions and his words—DNA can tell us what his true fate would have been had he not been killed by an assassin. Indeed, Lincoln's DNA, extracted from his remains, is an eternal text that need only be deciphered by contemporary molecular biologists.

As an immortal, historical text, DNA has also been called on to answer questions about geographical migrations and cultural interchange in the distant human past. The Human Genome Diversity Project is an international plan to use DNA from 500 distinct populations scattered around the world in an effort to understand human history. Blood samples containing DNA, to be collected from members of populations as diverse as the Yanomami of Venezuela and the Chukchi of northern Siberia, will be preserved and stored in a repository for future analysis. The project's promoters, most prominently the Stanford University population geneticist Luigi Luca Cavalli-Sforza, suggest that this collection of DNA can explain: the Bantu expansion in Africa, when the first agriculturalists appeared 2000 years ago; the origins of Native Americans and the timing and number of their migrations across the Bering Strait; and the relationships between linguistic groups around the world. Whether such questions can be answered by analysis of DNA has been questioned by critics of the project, including some anthropologists who played a role in planning it.

For us, the Genome Diversity Project is another example of the common construction of DNA as an immortal text, in this case a text in which human prehistory is written. In order to use comparisons of DNA to determine when and how human populations migrated across the Bering Strait, geneticists must make many assumptions about rates of change in DNA, geographical shifts, and early human culture. Like archaeologists they must work with fragmentary and incomplete evidence that cannot necessarily answer the questions put to it. Yet the range and ambition of the questions proposed suggests geneticists' faith in the molecular text as the "Bible" or "Book of Man," as well as their hopes that DNA can reveal even the most arcane truths of ancient human history.[42]

Richard Dawkins, in his popular 1976 book *The Selfish Gene*, called human beings "survival machines—robot vehicles that are blindly programmed to preserve the selfish molecules known as genes."[43] Dawkins may seem materialist and antireligious, but his extreme reductionism, in which the DNA appears as immortal and the individual body as ultimately irrelevant, is in many ways a theological narrative, resembling the belief that the things of this world (the body) do not matter, while the soul (DNA) lasts forever.

The immortality of DNA in Dawkins's account is grounded in reproductive processes: Genes live forever because they are replicated within living organisms on which they confer survival advantages. But DNA can be immortal in another sense: The molecule itself, in isolation from the living organism, can persist in fossilized form and, at least in fiction, make possible the retrieval of an organism long extinct. This is the basis of Michael Crichton's best-selling 1990 novel *Jurassic Park*, made by Steven Spielberg into a "blockbuster" film in 1993. The story is about corporate scientists who develop an island theme park filled with living dinosaurs. They produce the dinosaurs from bits of DNA extracted from dinosaur blood preserved within insects embedded in amber. The dinosaurs, however, turn out to be more aggressive and destructive than expected. Some of the smaller specimens escape on a supply boat and attack children on the mainland. In the final crisis a corrupt worker, bribed by a rival corporation, destroys the island's security system and is himself consumed by the island's *Tyrannosaurus rex*.

Crichton's plot drew on prevailing narratives of contemporary molecular biology in which DNA contains the complete instruction code for the living organism. While cloning a dinosaur is a theoretical possibility, there are serious practical problems with Crichton's scenario. None of the ancient

DNA thus far retrieved from amber has been complete; and DNA alone cannot make an organism. In all species, DNA interacts with its cellular environment—which includes maternally derived mitochondrial DNA specific to the species—to produce the developing embryo. No complete dinosaur cells survive, however, so that cloning a dinosaur would require the use of a cell from a species believed to be closely related to the dinosaur, such as the alligator. Cloning across species has never been successful, so it is not clear that dinosaurs could be produced even if an entire dinosaur genome were available. But in Crichton's story, if you want to get dinosaurs, all you need is dinosaur DNA. The powerful molecule with magical powers can resurrect the dead, even if the body in question has been dead for many millions of years.

Crichton's tale is a popular catechism promoting the idea of "immortal DNA." But it is also a morality play about forbidden fruit and the dangers of scientists playing God.

Forbidden Territory

In most cultures, some parts of the natural or social world are taboo. So too, DNA, in many stories, is a sacred territory, a taboo arena, that by virtue of its spiritual importance should never be manipulated. As the encoder of an essential self, a genetic soul, the genome has become forbidden ground.

The fear of tampering with genes is explicit in religious publications, where genes appear as "life's smallest components" and the "core" of humanness. In 1983, 21 Catholic bishops and a spectrum of other religious leaders wrote a widely disseminated statement demanding a ban on genetic engineering. Humans have no right to decide which genetic traits should be perpetuated, the statement declared; they have no right to "play God." Articles in religious magazines also express concerns about "tinkering" or "tampering" with genes. In 1989 a writer in the evangelical journal *Christianity Today* asked: "Is it permissible to alter humanness at its core, to tamper with our essential humanity? Genes are a core that should not be monkeyed with." An essay in the Jehovah's Witness publication *Plain Truth* questioned the hubris of contemporary genetics, which has made "man himself . . . the new God." Hundreds of years from now, the essay predicted, "humans . . . will look at our age and shake their heads in utter amazement. . . . They will wonder how we could possibly have believed that man alone was capable of solving his problems of disease. . . . The real God, not the one fashioned by man's religion and cloned in our image . . . will give us all the good things the genetic revolution promises."[44] To manipulate genes is to move them to the profane realm of engineering and technology. This, it is feared, will compromise their spiritual status. By opposing genetic engineering in these terms, such statements acknowledge the sacred power of DNA.

The sanctity of the genes is also a favorite subject in films about mutants, in science fiction novels, and in numerous revivals of the Frankenstein myth. All depict horrible consequences of genetic manipulation. In these stories, DNA is sacralized or forbidden territory, to be transgressed at a very high cost. They are traditional narratives of divine retribution for violating the sanctity of human life. But since the 1970s, they have appropriated the language of contemporary genetics.

A typical story appeared shortly after the 1976 controversy over recombinant DNA research.[45] Stephen R. Donaldson's *Animal Lover*, published in 1978, is about a geneticist, Avid Paracels, who becomes the victim of "genetic riots" that take place as the public became morally outraged by his efforts to develop a superior human being. Threatening the "sanctity of human life," the geneticist loses his grants and has to abandon his career. He is bitter: "By now I would have been making superman, . . . geniuses smart enough to run the country decently for a change, . . . a whole generation of people immune to disease." He plans his revenge and develops genetically altered animals capable of using advanced weapons, but he is thwarted by a cyborg who mortally wounds him in a climactic battle. "I can't understand why society tolerates mechanical monsters like you, but won't bear biological improvements," the dying scientist proclaims. "What's so sacred about biology?"[46]

In many subsequent science fiction novels, manipulating DNA leads to the creation of immoral or amoral human beings. The scientist in Robin Cook's *Mutation* (1989) produces a monster when he injects genes for intelligence into his own IVF-conceived baby son. The boy is indeed brilliant, but he is also cruel, emotionless, and totally manipulative.[47] In this and other stories, tampering with genes results in the dissolution of the family as the sacrosanct unit responsible for perpetuating the essential material.

Michael Stewart's *Prodigy* (1991) also features a geneticist who injects his IVF-conceived child with an extra intelligence gene. His wife, convinced of the importance of nurture, objects, saying that it is "just a short step from Mendel to Mengele." But the geneticist insists that "heredity is the dominant factor" and proceeds to manipulate their child's DNA without his wife's knowledge. Their daughter becomes an intellectual prodigy, but also a "living nightmare" with "evil built into her genes." The book's moral: "No man has the right to tamper with the building blocks of human life."[48]

The fear of tampering with genes is not limited to religious publications and science fiction plots. Science critics such as Jeremy Rifkin and some bioethicists express similar reservations. News reports about genetic manipulation also often dwell on potential dangers of genetic engineering. There are "perils" in "uncontrolled tampering," wrote a *Time* reporter. "Lurking behind every genetic dream come true is a possible *Brave New World* nightmare. . . . To unlock the secrets hidden in the chromosomes is to open up the question of who should play God with man's genes." An accompanying image portrayed scientists balancing on a tightrope of coiled DNA.[49] And an illustration for a *New York Times* article on gene therapy featured a drawing imitative of the famous Edvard Munch painting, *The Scream*. A figure stands, horrified, mouth agape, eyes wide open, its hair a mass of coiled DNA.[50]

This sacralization of DNA coexists in popular culture with another, contrasting view of DNA as utterly mechanistic and therefore dangerous in a different way. Some critics of the Human Genome Project express concerns that the ability to manipulate the genome through genetic engineering will desacralize the body by reducing it to a mechanistic entity. How, they ask, can we go on believing that the human body is sacred?[51] But we have observed a different set of images; the gene itself has been endowed with the qualities of a sacred object and the genome has become a fundamental text. In both the language of scientists and the parables of popular culture, the biological structure called DNA has assumed a nearly spiritual importance as a powerful and sacred object through which human life and fate can be explained and understood. Thus a January 1994 cover of *Time* depicts a man, arms extended in a Christlike pose; his torso, bathed in ethereal light, is inscribed with a double helix (see page 17). Such images give mystical and fantastic meaning to a molecular entity, and provide the foundation for the construct we call genetic essentialism.

Conveyed by scientists as they describe the meaning of their research, this idea of genetic essentialism has been readily adopted in popular forums where DNA—the invisible, eternal, and fundamental basis of human identity—has acquired many of the powers once granted to the immortal soul. Like the sacred texts of revealed religion, DNA explains our place in the world: our history, our social relationships, our behavior, our morality, and our fate.

4

The Molecular Family

A fertility handbook proclaims that "the desire for a family rises unbidden from our genetic souls." Infertile couples feel "uncontrollable devastation of their lives" when they cannot bear genetic children. And a genealogy search firm promises that "your genealogy will tell you what made you what you are today." In the 1990s genetic connections have come to define a new molecular family, one bound together less by history, tradition, or common experience than by shared DNA. The parent-child relationship, particularly, appears in popular culture as dependent less on caring contact than on genetic similarity, as adoptees search for genetic roots and prospective parents endure painful fertility treatments in order to produce a child who shares their DNA.

Ideas about what constitutes a relationship and what makes a relationship important or binding differ in different cultures. Kinship is a complicated cultural construction, and kinship ties are not necessarily drawn solely on the basis of biological relationships. Anthropological studies suggest kinship is "a system of cultural knowledge through which social practices are realized," rather than "a set of immutable biogenetic facts."[1] The anthropologist Marilyn Strathern notes that family bonds generally depend on both social agreements (for example, about who should live together) and biological relationships, but the relative importance of these two kinds of connections can vary greatly.[2]

In contemporary America, kinship is culturally defined as biological; the social aspects of kinship often seem invisible or unimportant, and the family commonly appears to be, as David Schneider has observed, a natural unit "based on the facts of nature."[3] Families that operate outside this model—in which there are adopted children or stepchildren, for example—must deal with a widespread social assessment that they are not "real." Such families exist and are recognized as families in law and in popular culture, but they are commonly perceived as unauthentic. Stories about the tragic fate of stepchildren—captured in the French fairy tale "Cinderella" but reiterated in contemporary stories of abused or abandoned stepchildren—express the cultural expectation that real families are based on ties of blood.

Just as the nuclear family is less inclusive than the larger extended family, so the new molecular family is less inclusive than the nuclear family, for the molecular family centers on the dyad of biological parent and child. The defining relationship of the molecular family is between genetic parent and child; all others are weak in comparison. This constricted vision of human bonds—this narrowing definition of connection and relation—may be a response to the changed social meaning of family life in the 1990s. Alternative partnerships, working mothers, and high divorce rates appear in popular culture as troubling problems. In fact, divorce and working mothers have been common for much of human history: the golden age of the family probably never occurred. Yet today the family seems to be in a special state of crisis, threatened by feminism, divorce, the gay rights movement, the ability of children to sue their parents, the complex arrangements enabled by new reproductive technologies, and other social changes. As an apparently besieged institution, the family is a focus of public concern, and this is reflected in news reports, television sitcoms and family magazines.[4]

The notion of the molecular family is based on the cultural expectation that a biological entity can determine emotional connections and social bonds—that genetics can link us to each other and somehow preserve a reliable model for a family. Since it is beyond culture, outside of time, DNA seems to be of durable or permanent significance. Genetic ties seem to ground family relationships in a stable and well-defined unit, providing the individual with indisputable roots more reliable than the ephemeral ties of love, friendship, marital vows, or shared values. In popular theories of evolutionary psychology, the idea that genetic bonds are truly lasting is used to naturalize divorce and infidelity through claims that marriage partners have valid biological reasons to desert each other.

In such accounts, genes create ties that can (or should) never be severed and validate the individual as genetically placed in an unambiguous relationship to others. Yet rather than helping to resolve tensions about the family, such expectations exacerbate them, for if genetic similarity is crucial to parent-child relationships and to personal identity, then the realities of infertility and adoption are threatening to the individuals and families concerned and to the strength of the family. And if family relationships depend on shared DNA, then to be raised by a "biological stranger"[5]—in the words of a court in a child custody case—is a dislocating and isolating experience.

The concept of the molecular family appears in tales of infertility, adoption, reproductive choice, and the search for biological "roots." These are often stories of desperation:[6] the adopted child seeks his parents out of a profound need to connect himself to a biologically related parent; the infertile couple agree to surrogacy out of a consuming need to have a baby who carries their genes; the aging baby boomer be-

comes a single mother out of the desperate drive to perpetuate her genes; and individuals track their family genealogies
as clues to both social and biological identity.

The Desperation of Infertility

For the generation born from 1949 to 1961, the 80s and 90s
are the time of infant care and "cocooning." For women who
have not borne children, images of the ticking "biological
clock" and "motherhood deferred" construct reproductive
potential as central to identity. In women's magazines, television programming, and films, a woman's need to reproduce
appears as overwhelming and the experience of infertility as
self-annihilating.

There are, of course, many reasons to want a child and
many social pressures encouraging reproduction. But in
popular narratives, the drive to bear children is often
equated with a drive to perpetuate DNA and a need for
genetic continuity. This message emerges in stories of
women who, by pursuing careers and postponing pregnancy,
are "running out of time." In the film *Look Who's Talking*,
Molly, a successful 33-year-old accountant who becomes
pregnant, goes to a doctor to ask about abortion and is
warned that "her biological clock is ticking." That night,
Molly dreams that she is hanging from a ticking clock on the
upper floors of a skyscraper and she falls to her (genetic)
death. She decides to bear the child.[7] In this and many other
stories of the ticking biological clock, postponing motherhood is a fundamental threat to survival, for it may preclude
the creation of genetic bonds.[8]

Such stories can be used in critiques of feminism. Anne
Taylor Fleming's 1994 biographical account of infertility,
Motherhood Deferred, describes her experiences as a partici-
pant in biomedical infertility research and her anguish at her
inability to conceive a child. Fleming has concluded that she
fell under the spell of the "angry, childless and unmarried
ideologues" of the feminist movement and that putting work
ahead of motherhood has made her and other women seek-

ing infertility treatment into "modern day Hester Prynnes," women who must wear a "scarlet letter" identifying themselves as inadequate.[9]

The narrative is hardly a new one, but its current construction reflects changes both in social practice and in contemporary biomedical science. In the late nineteenth century, physicians feared that the higher education of women (by then a significant social reality) would lead to the degeneration of their reproductive organs. They pictured the female body as a closed system with only a limited amount of "vital force": the brain and the ovaries could not develop at the same time.[10] Similarly, in the contemporary debate, women who pursue intellectual or professional interests and thereby delay childbearing are threatened with reproductive difficulties. The problem is now conceptualized in terms of recent scientific understanding of reproductive biology and the statistical link of infertility with maternal age. But the underlying notion—that intellectual work is antithetical to female reproduction—is consistent, as is the connection between reproductive ability and self-validation. A woman who cannot bear children is, in the infertility literature and beyond, profoundly impaired.

Many cultures, of course, define women in terms of their reproductive ability. In Bantu society, for example, perpetuation of the community through the creation of family underlies the definition of personhood, so that women who are unable to give birth are considered to be nonpersons.[11] But in contemporary American culture, the pain of infertility is often located at the molecular level, defined as a basic need for genetic identity, continuity, and immortality. Genetic children thus become critical possessions. An adopted child may provide the experience of being a parent; in many descriptions, however, adoption fails because it cannot provide the biological connection upon which family and self-concept are assumed to be based.

A popular book about infertility refers to the "biological handicap" of those unable to bear children. The author interviews infertile couples who feel abnormal and diseased because they cannot bear genetic children.[12] A fertility handbook conflates personal fulfillment, religious impulse, and genetic destiny to explain the urgency to reproduce: "Call it a cosmic spark or spiritual fulfilment, biological need or human destiny, the desire for a family rises unbidden from our genetic souls."[13] A character undergoing fertility treatment in a women's magazine short story says she and her husband are "like warriors going into battle" against a body that has "betrayed" her.[14] And in *Resolve*, a newsletter on infertility, a writer describes the feelings of inadequacy and the lack of self-esteem among women unable to conceive. Having a baby is essential if a woman is to be whole, for the experiences of pregnancy and motherhood are "the core of women's being."[15]

Popular journalism reports the dramatic sacrifices of couples who suffer from infertility and who "try everything," regardless of cost, in the attempt to have a child. Stock phrases—"baby craving," "baby fever," and "the quiet pain of infertility"—express the critical importance attached to having a genetically related child.[16] Infertile couples are described as having lost out in nature's lottery. "Like desperate gamblers going for the pot each time . . . infertile people wait for the miracle that could happen each month and doesn't." There is a "special hell of infertility" and "demons that eat at the soul."[17]

Infertility treatment has become big business.[18] More than $1 billion is spent annually on these treatments, with more than 1 million patients seeking such procedures as in-vitro fertilization, gamete intra-Fallopian transfer, frozen embryo transfer, and other technological solutions. Although requests for such treatments partly reflect the shortage of babies to adopt since the legalization of abortion in 1973, demands have increased remarkably in recent years. In 1987, there were fewer than 50 IVF clinics in the United States. In 1992, there were 235. New reproductive technologies, suggests a *New York Times* writer, "have spun a world in which every gamete gone astray becomes a phantom baby."[19]

In a consumer culture, it is not surprising that a market should develop around reproduction; indeed, the language of infertility is replete with commercial images—banks, deposits, fees for services, property, products, and possessions. But as such language suggests, reproductive interventions are costly (they are rarely covered by medical insurance). They can also be painful, humiliating, dangerous to the woman's health, extremely time-consuming, and frustrating. At 10 to 15 percent, the success rate of in vitro fertilization is poor. Critics have questioned the allocation of research funding for infertility in the midst of a worldwide population boom. Yet stories of $50,000 babies and their beaming parents support the idea that "any struggle is worth a shot at this prize."[20] The prize is not merely a child—for a family can have a child through adoption—but a genetically related child, produced through the experience of pregnancy.

Pregnancy in this literature, however, is more than a biological experience. It often appears as a validation of self and a source of identity for women who, by bearing a child, forge an unbreakable genetic bond. The image of actress Demi Moore, nine months pregnant and nude on the cover of *Vanity Fair* in August 1991, was a powerful expression of this celebratory view of pregnancy. A journalist, perhaps more down to earth, described her own pregnancy as both unpleasant and socially rewarding: "Even though I hated the sweating, the heartburn and the funny underpants I wore when I was pregnant, I liked the feeling that being pregnant was something. SOMETHING! I stuck out and waddled, and society smiled at me and gave me seats on busses . . . I felt queenly and grateful and worthy of being served."[21] Both actress and journalist celebrated the experience as a validation of their essential worth. Yet even a vicarious pregnancy through surrogacy appears to be more acceptable, to many prospective parents, than the adoption of a "biological stranger."

Surrogacy is, in effect, a social manifestation of genetic essentialism. The practice itself has been viewed with some suspicion in the wake of widely publicized legal disputes over the custody of the infants. But many popular magazines have suggested that the genetic imperative—the need to create genetic ties—legitimates surrogacy, despite its problems. According to *Glamour*, "surrogacy may soon be as American as baseball."[22] The story described a couple who had their first baby by using the husband's sperm and a surrogate's egg, and their second by implanting the wife's fertilized embryo into a surrogate mother. "This way, Linda could have a baby who was genetically hers as well."

There are several types of surrogacy. A surrogate mother can simply be inseminated with the sperm of a man whose wife is unable to bear a child; or techniques of *in vitro fertilization* can be used to transfer a fertilized embryo from a couple to a surrogate who would then bear the child.

In either case, stories of surrogacy trivialize the surrogate's bond to the baby—a bond which may or may not be genetic, but which is always indisputably biological. Fetal environment, which has at times appeared critical to child development, is presented in the surrogacy debate as almost irrelevant. In the 1950s, for example, fetal environment was considered to have a deterministic influence on a child's temperament. A *Parents* magazine article described the prevailing scientific view: "What is being realized today is that the processes that go on within the tiny organisms in the first few months . . . are vital. They set a pattern which may make or break the grownup."[23] And more recently in debates over occupational health, fetal protection looms large as a justification for excluding women from many jobs.[24] Yet in the surrogacy literature, the fetal environment provided by the surrogate appears as a neutral incubator, a space in which the fetus (the packet of genes) can grow, but which hardly engages the surrogate in a relationship with the child. The surrogate may have nurtured the fetus and carried the baby to term, but she seems to have no legitimate maternal interest in its welfare. Without the genetic relationship—the shared DNA—a woman has no proper claim as a mother.

The growing practice of surrogacy has moved from the newspaper headlines to the soaps. Soap opera stories about such arrangements build on the importance of genetic relationships for purposes of inheritance and often revolve around immense fortunes. For example, in "As the World Turns," Dana became a surrogate mother (a "hired womb") for the wealthy Crawford family, which needed an heir and provided both egg and sperm.[25] The baby she carried was essentially a commercial product, the genetic property of the Crawford family, to be produced by Dana's body so that the Crawfords would have someone to whom they could leave their money. The man who had provided the sperm reminded Dana that she must not become attached to the baby because she was not genetically related. Then Dana's exhusband kidnapped her, demanding 2 million dollars' ransom for her and the baby. Money and greed, pervading stories of surrogacy in the soaps, make explicit the idea of DNA as valuable familial property.

If infertility appears in popular culture as a devastating loss of the genetic bond, so the experience of being adopted appears as a devastating barrier to the development of a confident self-identity. Those raised by parents who are not biologically related now commonly confront popular expectations that they must search for their genetic roots. Some describe their quests in confessional essays or interviews as necessary, even crucial, because personal identity depends on knowing one's "real" parents.

The Search for Genetic "Roots"

The increasing popular interest in biological relationships appears in accounts of adoption and in stories about adoptees who are searching for their biological roots. The message is reinforced by visual images; for example, a drawing in *Esquire* of a man surrounded by the powerful roots of a large tree shows each root to be the ladder of the double helix.[26]

In 1979 Betty Jean Lifton wrote a best-selling book on "the adoption experience," laying out the problems of identity experienced by adoptees and urging them to search for the "missing pieces" of their lives. Lifton's 1994 book on the adoptees' "quest for wholeness" refers to adopted children as "fantasy people" who are "invisible." There is, she says, "a hole at the center of their being." They are "people without selves" because they are cut off from their biological connections, their "soul mates," and they "don't know what genes they got."[27]

Her message has been amplified by magazine stories suggesting that finding one's biological parents is essential to

becoming a functional adult, an integrated self. News articles and television talk shows feature emotional public confessions by adoptees who express their need to locate genetic connections.[28] In *Newsweek,* a reporter interviews an adoption counselor who describes a client's search for her biological parents as a quest for "wholeness."[29] A *New York Times* writer observes: "For the adopted, the issue of roots grows stronger."[30]

Popular images of adoptee quests exaggerate the reality. In a study of TV dramas, George Gerbner, a media analyst at the University of Pennsylvania, found that one in five adopted television characters chooses to track down birth parents. This percentage is 18 times greater than in real life. And on TV, one in three adoptees expresses worry that birth parents will eventually try to find him or her, but in fact such efforts by birth parents are extremely rare.[31]

Why does the saga of the lost connection—of the costly search for genetic relations—appear so frequently in popular fiction and film? What are the archetypal elements of this story that make it relevant to a larger audience? We suggest that the formulaic message of such stories is that of genetic essentialism; they appeal because they confirm a widespread cultural expectation that identity is somehow linked to shared genes.

This was the theme of Mona Simpson's 1991 novel *The Lost Father*, featuring a student's lifelong quest to find her father—a "search for the man without whom she could never find herself," according to the book jacket promotion.[32] A 1991 off-Broadway play, *Phantasie*, portrayed a 35-year-old woman who dreamt about her birth mother and the circumstances in which she was given away. Finding her mother, she said, was finding herself.[33] And in one episode of the primetime cartoon show "The Simpsons," Homer Simpson searched for his long lost brother, who had been adopted at birth, because "it's my life." The brother, finally found, was wealthy but, despite his money, miserable. The reason? He lacked family connections.[34]

Stories about quests for lost biological connections often present them as a search for self. A journalist's confessional account of his lengthy and agonizing search for his biological mother followed a format common to the "holy quest" as a transformative journey to seek personal identity. The journalist described himself as neither "real" nor "whole" without knowing his biological mother: he would never be complete without finding this "missing link." He therefore pursued her so relentlessly that the adoption agency threatened to have him arrested for harassment. For twelve years he continued, overcoming one obstacle after another, but finally gave up the search. Then his fortune changed. The agency called to tell him that they had located his mother. "Goose bumps covered my spine, legs, head, as if I was having a stroke." It took only a few words with his mother to recognize in her his own personality. When he met her, "her face was eerily familiar. I had been looking at it in the mirror my whole life . . . I was real."

The next month's issue of the magazine continued the story of his "unquenching thirst" to find his roots. This time he described his meeting with his genetic father: "I sensed a kind of completion. . . . For the first time in my life I could stop searching."[35]

A provocative PBS documentary in the fall of 1992, however, recorded one of the possible consequences of such searches. A young filmmaker told the story of his search for the biological father whom he had never met. He began this search out of a confused and inchoate sense of personal need. His mother dodged his questions, telling him only that he was conceived as the result of a brief affair. He eventually found his father and expected that meeting him would reveal something about himself—that it would resolve questions about his identity. But this was not borne out. The filmmaker's expectations, reflecting the cultural importance we attach to genetic relationships, were dashed, and the film ended not on a note of triumph but of profound disappointment.[36]

Even if the result is unsatisfying, the quest can be portrayed as inevitable. On a CBS docudrama, "Family Secret," an adopted boy raised in a stable and caring family had always felt lost—"without a home." He eventually tracked down his genetic mother. But his mother, now raising four other children, did not want to see him; she associated him with a bitter incident in her past (this boy was the product of a rape many years earlier). Her refusal to see him was a devastating blow, and the boy, years later, became a rapist himself. The story did not directly explain the cause of his criminal behavior, but it emphasized the importance of genetic bonds to adoptees, who may suffer terrible consequences if such bonds cannot be forged.[37] Those who lack genetic connections, who do not know their roots, appear in popular culture as damaged, devastated, or even doomed.

The idea that an essential element of identity is "missing" shapes these stories about the quest for roots. They suggest that to find one's genetic parents is to find one's essence, and that without this essence a person can never be complete, whole, fulfilled. As one man put it: "As soon as I searched and found the information . . . I felt more worthwhile in the world—as though I belonged better. Beforehand, a part of me had always been missing."[38] A *Psychology Today* article explored the problem of "genealogical bewilderment"—the idea that adoptees suffer lower self esteem because they do not know their genetic roots.[39] And a 1993 biography of Walt Disney portrayed him as "haunted by family tensions, addled by drink and pills," because he was an adopted child who never knew his biological parents.[40]

The preoccupation with genetic relationships can stigmatize the experience of adoption: "Adoption is just not the same as having a biological child," said a woman interviewed in an infertility book.[41] While some stigmas of adoption have disappeared—the "sin" of illegitimacy, for example—new stigmas have emerged: "Couples considering adoption are often warned that they're asking for aggravation and told of adopted children who 'just didn't work out' for one reason or another."[42] Adopted children are often regarded as high risk commodities, though a major study found that, contrary to popular belief, adopted teenagers are no more likely than other adolescents to suffer from mental health or identity problems.[42] In *Parents* magazine, a woman with two adopted girls wrote about her neighbors' responses. One warned her about "damaged goods." Another said, "I'd like to adopt but my husband would never approve. He said if he can't have his own, he doesn't want someone else's leftovers."[44]

Fears about inherited flaws are compounded by anxieties about genetic continuity. A woman who had adopted several children said she was plagued by a certain sadness: "We cannot look for my eyes or Jay's nose in our daughter's face. We can't hope that Amy has inherited her father's singing ability, or her uncle's keen analytic mind." There is "a stab of sadness that my father does not live though my adopted daughter."[45] In an interview in *Glamour,* a man whose wife cannot bear children explains his reasons for not wanting to adopt. "I come from a very old family. . . . I want to carry on the family genes." He and his wife hired a surrogate mother who agreed to be implanted, through IVF, with their fertilized embryo.[46]

In a 1993 book that is highly critical of the stigmatization of adoption, Elizabeth Bartholet described, from her own experience, the "regulatory obstacle course" faced by those who wish to adopt. Adoption, she argued, is seen as a "desperate last resort," a "debased form of parenting," when it should be "a positive alternative to biological parenting."[47] Adoptive parents are asked about their children's "real" or "natural" parents, as if those who nurture children are unnatural substitutes. "Parenting is equated with procreation and kinship with blood links. It is only genetically linked parents who are truly entitled to possess their children."[48] Bartholet's observations reflect in part the emergence of a vocal anti-adoption movement over the last decade.

Groups opposing the practice of adoption have proliferated. They label adoption a "pathology," arguing that genetics is the basis of identity, and that adoptees are "amputees." They circulate pamphlets with such titles as "Death By Adoption."[49] And just as infertility has generated a flourishing industry, so too has the need for genetic connections; commercial firms and social services, ready to assist in post-adoption searches, have proliferated. A directory of national social service organizations lists 33 national groups serving adoptees, many with local branches. Fifteen of these have been organized since 1980. Half of them advertise that they are engaged in searches; their names, such as Origins, Operation Identity, Adoptees Liberty Movement, Concerned United Birthparents, and Reunite, suggest their role. The International Soundex Reunion Registry, affiliated with 450 groups worldwide, operates a registry for those seeking blood relatives. It keeps a genetic data bank recording the medical and genetic data of its registrants.[50]

Search groups are sustained by their belief in "bonding by blood"—the critical importance of knowing genetic roots. The president of the Adoptees Liberty Movement observed that "The adopted infant is no blank slate. He brings his chromosomes with him." From this she concluded that the supposition underlying adoption procedures, that "love and other expenditures should enable the adoptive parents to expunge his heredity" is wrong: "The destruction of the heredity and identity of the adopted person is a deprivation for which there is no compensation."[51]

Adoptive search groups have generated a popular movement, attracting constituents by promising redemption through reunion. They promulgate their beliefs through television stories, personal accounts by clients, and visual images that convey the message of genetic essentialism: those who lack genetic connections are incomplete. Their efforts have been reflected in legislative proposals to change adoption procedures so as to facilitate searches: Proposed legislation in New York State, for example, is based on the belief, stated by Governor Mario Cuomo's wife, that "adoptees have the same right all the rest of us have: to know our identity, to know our roots, and to know ourselves."[52]

Embedded in these images of adoption is a broader social agenda—to find a cure for the troubled family, to enhance so-called family values. Genes, they suggest, link people securely to each other, grounding family ties in a powerful and unambiguous biological entity, DNA. Yet only the dyad of parent and child are linked through DNA. Indeed evolutionary psychologists, noting the potential genetic advantage (especially for males) of seeking multiple partners who will continue to reproduce, offer "natural" explanations for the instabilities of contemporary family life.[53]

The DNA Family

The family drama has long been a television mainstay, and the families portrayed in prime time television programs have changed with their audiences. Historian George Lipsitz, in *Time Passages*, and sociologist Ella Taylor, in *Prime Time Families*, have described the successful formulae that shaped popular family series such as the long-running "I Remember Mama" and "The Goldbergs."[54] These early programs, aired in the 1940s, portrayed working class households in which several generations lived together. They were succeeded by sitcoms of the 1950s and 1960s, in which happy middle-class, suburban families made decisions about money and consumption; their possessions were the key to family stability. In the 1970s, prime-time family programs began to portray "trouble": urban unrest; divorce; race, sex, and social-class tensions. Anxieties about changes in family life "converted the domestic sphere into a febrile and argumentative war zone" in such popular shows as "All in the Family" and "Archie Bunker's Place."[55] By the late 1980s these themes had given way to an ideology of privatism: television dramas reflected reduced interest in public life and growing concerns that the moral authority of the family was declining.

At the same time, other programs began to feature unconventional family arrangements. As noted by Taylor, there was the all-male family ("Dads"; "My Two Dads"), the single mother ("Murphy Brown"; "Kate and Allie"), the all female household ("The Golden Girls"; "Designing Women"), and the mixed-race family ("Diff'rent Strokes"). These programs articulated social expectations about family life and explored the basis of enduring family relationships. Their characters sometimes expressed a sense of discomfort at their unconventional living arrangements. Families held together by emotional ties or shared experiences seemed fragile and insecure. "Do I have to be a relative to be family?" asked a confused boy in the series "Who's the Boss." "Not necessarily," said his mother. "A family means people who share each other's lives and care about each other."[56]

Amidst this confusion and concern about the nature and definition of the family, DNA testing has appeared as a way to clarify relationships. In sitcoms and soaps portraying "odd" family situations, DNA testing has become a way for families, never quite at ease with their nongenetic ties, to establish "real" (that is, genetic) relationships. For example, in NBC's "Pop the Question" two men, Michael and Joey, were comfortably raising a child named Nicole until they felt compelled to discover which of them was her biological father. To settle the matter, they decided to have genetic tests. But when the results came in, they did not want to open the envelope. Ultimately, they decided that the real father was the one who was actually raising Nicole; since they were both filling that role, they were both her father. The concept of genetic connections had challenged their assumptions about the importance of shared experience and the best interests of the child, but in this case, social bonds prevailed.[57]

In other television programming, characters employ DNA testing to ascertain their family ties—sometimes as a way to establish the rights of inheritance, sometimes to discover the meaning of their life. "My whole life is a lie," said Nick in a "Guiding Light" episode when he found out the truth about his biological parentage.[58] In "All My Children" Haley, a teenager, lived happily with her uncle Trevor until the day she discovered that another man was her biological father. She moved out of her uncle's home to live with her biological father (and to inherit his millions).[59]

The interest in familial genetic ties is also reflected in the extensive media coverage of sperm bank mix-ups and new-born babies switched in the hospital nursery. Such themes have historical resonance: baby switching stories in the classic mode of *Oliver Twist* have long been a way to suggest the natural basis of social class and economic privilege. They continue to fascinate the media, though the message today is less explicitly about social class than about the relative importance of social and genetic relationships. For example, the discovery in 1989 of a baby mix-up error in a Florida hospital that had occurred ten years earlier generated a deluge of news reports, a television mini-series, a documentary, and many magazine stories. In 1979, the Mays family went home with Kimberly, and the Twiggs family with Arline. Arline was a sickly child, and in the midst of her extensive medical care, Mr. and Mrs. Twigg discovered that she was not their biological child. After Arline died, they sought to find their daughter and claim custody of her. They eventually settled for visiting rights—Kimberly was then happily living with the father who raised her.

Media coverage of the case focused on the struggle between genetic and emotional bonds. In an NBC docu-drama based on the case, an actor portraying Mr. Twigg displayed a photograph of his large extended family. He explained that he wanted his genetic daughter to have the benefit of that family connection: She belonged in the picture.[60] The people in the family photograph were strangers to his daughter (who was, by then, a teenager). Yet, by virtue of her DNA, they seemed to have a special claim on her. The lawyer for the Twiggs insisted on the importance of "pure biology . . . as opposed to a fully developed parental relationship with adults who were not the natural parents."[61] At the center of all this publicity was a young woman who apparently had mixed feelings about both sets of parents. In 1993, Kimberly formally severed all ties with the Twiggs and announced her plans to stay permanently with the man with whom she had grown up. But less than a year later, in May 1994, she changed her mind and moved in with her genetic parents, who were then awarded legal custody.

The media coverage of child custody disputes also reflects the tension over the relative importance of nature and nurture, of genetic and social ties. The prolonged, confusing, and highly publicized battle over the custody of two-year-old Jessica DeBoer is a case in point. Jessica, born in Iowa, was given up at birth with the consent of her biological mother and a man whom the mother identified as the biological father. The child was placed immediately with the DeBoer family of Michigan, but within ten days of giving up the child, the biological mother changed her mind. She married Daniel Schmidt, who used a DNA test to prove that he was the biological father and asserted that he had never given consent for adoption. Their early demands for Jessica's return prevented a legal adoption, but the interstate court battle lasted two years, during which Jessica lived with the DeBoers. While the outcome of the case turned less on biology than on legal precedent, the media focused on the conflict between the child's emotional relationship with her adoptive parents—the only parents she had ever known—and the genetic claims of the birth parents. Reporters cited experts on the importance of emotional bonds and experts on the need for biological connections. Parenthood, according to one press account, is widely understood to be a biological classification.[62] But another insisted that "biology doesn't make the parent."[63]

Similar tensions were expressed in the media coverage of a 1992 custody dispute involving the separation of siblings.[64] The adoptive parents of one of the children, Michael, sued a social service department to overturn the adoption of

Michael's sister by another family. By the time Michael's parents brought suit, however, the sister had lived with the other family for more than two years. A *New York Times* article quoted an expert from an adoption program who stressed the importance of keeping siblings together: "These children are often deprived of one of the last remaining human connections really meaningful in their lives." Michael's sister, however, defined her own meaningful connections in quite different terms. Asked by her brother to move away from her adoptive parents, she responded: "No, I'm going to stay with my family."

The American preoccupation with the integrity of the DNA family is also apparent in the growing popularity of genealogical research. Genealogy has become a major hobby. The interest in family trees, of course, predates the preoccupation with biological roots and, indeed, began with the impulse to preserve family traditions. But in recent years genealogical services have proliferated, as have the number of genealogy books. The listings under "genealogy" in *Books in Print* increased from 115 in 1980–1981 to 210 in 1988–1989 and 300 in 1992–1993. They include bibliographies, source books, and books for children; the Christmas season always generates a series of gift volumes for children with titles like "Generations," "The Great American Ancestor Hunt," and "Pursuing the Past." A computer program called Family Tree Maker sells for $49.95.[65] Solicitations to subscribe to genealogy services arrive in the mail: "Nelkins Across America" and "Nelkins Since the Civil War."

Genealogies can be a way to reclaim history and to bring historical traditions to bear on contemporary culture. Or genealogical searches can be a means to establish personal identity on the basis of genetic connections. The current preoccupation with "family trees" suggests the latter: knowing one's heredity is valued as a means of knowing oneself. A genealogy research service advertises: "Do you ever wonder about what makes you different from your best friend? . . . Your genealogy will tell you what made you what you are today."[66] An article in *Forbes* describes the increasing number of "hereditary groupies" who measure their worth by their ancestral roots.[67] The Daughters of the American Revolution have developed a Family Tree Genetic Directory, inspiring one journalist to quip about the DNA of the DAR and to predict a whole new field of science called "club genetics," bent on proving that a person's genes can be determined by which clubs they joined.[68] But the message conveyed by most genealogy books is the permanence and stability of genetic ties. *Parenting* magazine advises its readers to help their children enhance their sense of identity and security by building family trees; unlike other relationships, the ones charted in the family tree cannot be broken.[69]

Genealogy searches, tales of baby switching, and the questions raised about adoption have used genetic explanations to support the idea of family values. They assume that families must be defined in terms of biological connections. But such connections hardly assure a stable and nurturant family environment. Thus, genetic explanations are also used to explain the "demise of monogamy," and the unstable realities of contemporary family life. A 1994 *Time* cover pictures a fractured wedding band captioned: "Infidelity: It May be in our Genes."[70]

Actual family units are sustained, however, more by social forces than biological realities. Many professionals define the family as a social unit that may include relationships based on emotional connection and commitment rather than on biological ties.[71] Indeed, more than 15.5 million children in America do not live with their two biological parents.[72] And millions of families function well on the basis of a social contract that has little to do with the impulses of the genes.

But the social realities of divorce and remarriage that threaten the conventional nuclear family strongly reinforce the notion of the molecular family, which logically includes only persons linked by DNA. Because they do not share genetic material, marital partners are, according to evolutionary psychologists, unlikely to stay together. Like sociobiologists, they argue that social behavior can be understood as the consequence of genetic forces. David Buss, a University of Michigan scientist who theorizes about the biological reasons for human emotions, believes that infidelity and divorce are natural because individuals have a genetic stake in multiple partners: men because they hope to father more children; women because a sexual liaison may convince a man to care for or support a child who is not actually his own. Romantic relationships, claims journalist Robert Wright, are grounded in the "strategies of the genes"; and the roller coaster emotions accompanying romance are "the handiwork of natural selection" remaining with us today because in the past they led to behaviors that helped to spread genes.[73] This popular construction of romance and marriage explains family instabilities as the natural consequence of our programming for genetic survival. The only real (that is, genetically significant) relationship is that between parent and child.

Popular interest in genetic connections coincides with the increasing visibility of—and public discomfort with—unconventional family arrangements. Genes as the basis for family stability are particularly appealing in a society reacting with anxiety to the ambiguities of "new families" and "blended families," of homosexuality, single motherhood, working mothers, and assorted "alternative lifestyles." Narratives of popular culture portray the molecular family as a stable, "natural" unit at a time when families appear to be chronically unstable and highly fragile. Just as these narratives portray genetics as the basis of proper family relationships and the source of individual identity, so too they construct the individual as a biologically determined being whose health and illness, behavior and intelligence, success and failure are all dictated by genes.

5

Elvis's DNA

In popular culture, Elvis Presley has become a genetic construct, driven by his genes to his unlikely destiny. In a 1985 biography, for example, Elaine Dundy attributed Presley's success to the genetic characteristics of his mother's multiethnic family: "Genetically speaking," she wrote, "what produced Elvis was quite a mixture." To his "French Norman blood was added Scots-Irish blood," as well as "the Indian strain supplying the mystery and the Jewish strain supplying spectacular showmanship." All this, combined with his "circumstances, social conditioning, and religious upbringing . . . [produced] the enigma that was Elvis."[1] Dundy traced Elvis's musical talents to his father (who "had a very good voice") and his mother (who had "the instincts of a performer"). His parents did provide a musical environment, Dundy noted, but "even without it, one wonders if Elvis, with his biological musical equipment would not still have become a virtuoso."[2]

Another Elvis biographer, Albert Goldman, focused on his subject's "bad" genes, describing him as "the victim of a fatal hereditary disposition."[3] Using language reminiscent of the stories of the Jukeses and the Kallikaks, the degenerate families of the early eugenics movement, Goldman attributed Elvis's character to ancestors who constituted "a distinctive breed of southern yeomanry" commonly known as hillbillies. A genealogy research organization, Goldman said, had traced Presley's lineage back nine generations to a nineteenth-century "coward, deserter and bigamist." In Goldman's narrative, this genetic heritage explained Elvis's downfall: his addiction to drugs and alcohol, his emotional disorders, and his premature death were all in his genes. His fate was a readout of his DNA.

The idea that "good" and "bad" character traits (and destinies) are the consequence of "good" and "bad" genes appears in a wide range of popular sources. In these works the gene is described in moral terms, and it seems to dictate the actions of criminals, celebrities, political leaders, and literary and scientific figures. Films present stories of "tainted blood," and "born achievers," of success and failure, of kindness and cruelty, all written in the genes. The most complicated human traits are also blamed on DNA. Media sources feature jokes about Republican genes, MBA genes, lawyer genes, and public interest genes.[4] Human behaviors linked to DNA in these accounts range from the trivial—a preference for flashy belt buckles—to the tragic—a desire to murder children.

Such popular constructions of behavior draw on the increasing public legitimacy of the scientific field of behavioral genetics. Behavioral geneticists have been able to demonstrate that some relatively complicated behaviors—certainly in experimental animals and possibly in human beings—are genetically determined.[5] Studies of animals reveal the genetic bases of survival instincts, mating rituals, and certain aspects of learning and memory. Border collies herd sheep in a unique, characteristic way whether they have been trained or not, even if they have never seen sheep before. Some behaviors associated with particular hormones have been indirectly linked to genes: Both aggressive and nurturing behaviors—in mice—can be manipulated with adjustments of hormone levels. Though such research highlights the biological events involved in some behaviors, it does not support the popular idea that genes determine human personality traits or such complex phenomena as success, failure, political leanings, or criminality.

Nonetheless, the claims that genes control human behaviors have received significant support from some behavioral geneticists who have positioned themselves as public scientists. Among the most cited and widely promoted scientists in this field is University of Minnesota psychologist Thomas Bouchard. Bouchard, a student of Arthur Jensen, has studied identical twins reared apart in order to determine the relationship between genetics and IQ, personality, and behavior. Bouchard's work has attracted significant popular attention since he began promoting his findings in 1982, but it has been controversial in the scientific community. Identical twins growing up in different families have long been seen as "natural experiments" in human genetics, even by the eugenicists of the 1920s (see Chapter 2). Bouchard, like others before him, has concluded that all similarities in identical twins reared apart are caused by their shared genes. But Bouchard's research subjects were self-selected (he advertised to find them) and interested in being twinlike. Some of them had also been reared together for several years before they were adopted into different families, therefore sharing at least an early environment. In addition, in any population a certain number of similarities will appear by chance. The fact that two people enjoy the same soft drink— in a culture in which soft drinks are widely consumed—is not evidence that they share a gene for the consumption of that soft drink.

For years Bouchard had problems getting his papers accepted for publication in scientific journals. Convinced of his work's importance, however, he submitted his findings to the press before they were peer reviewed, or even when they had been rejected by scientific publications.[6] The media responded with extraordinary interest, attracted to the drama in "the eerie world of re-united twins" and the potential for controversy over the sensitive issue of genes and IQ. *U.S. News and World Report* reported on the twin studies by describing the character traits that are "bred in our bones." Quirks such as wearing flashy belt buckles, liking particular television programs, or drinking coffee cold, problems such as addiction or eating disorders, were all described as originating in the genes.[7] The "Donahue Show" began a program on the twin studies with films on animal behavior, suggesting that, like animals, we "get a push before the womb." *Time* magazine criticized the political liberals who explained crime and poverty as byproducts of destructive environments. The twin studies were "one more proof that parenting has its limits."[8] The *Boston Globe* announced that "geneticists now have ascendancy in the nature-nurture debate."[9]

In October 1990, *Science* became the first major professional journal to publish Bouchard's work.[10] There followed a media blitz. The *Philadelphia Inquirer* headlined its front-page story: "Personality mostly a matter of genes" and welcomed the "landmark" study that proved that personality is put in place at the "instant" of conception. Even religiosity and church attendance, the article said, were determined by genes.[11] Magazine articles touted Bouchard's research as part of the swelling tide of evidence for the importance of genes.

Since 1983, when behavioral genetics first appeared as a category in the *Reader's Guide to Periodical Literature*, hundreds of articles about the relationship between genetics and behavior have appeared in magazines, newspapers, and fictional accounts, often presented as the cutting edge of current science. Included among the traits attributed to heredity have been mental illness, aggression, homosexuality, exhibitionism, dyslexia, addiction, job and educational success, arson, tendency to tease, propensity for risk taking, timidity, social potency, tendency to giggle or to use hurtful words, traditionalism, and zest for life.[12]

Many of the stories of good and bad traits address a common and troubling contradiction. Why do some individuals, despite extremely difficult childhoods, become productive, even celebrated members of society, while other children, granted every opportunity and advantage, turn out badly? What accounts for the frequent disparity between achievement and hard work? Genetics appears to provide an explanation. Individuals succeed or fail not so much because of their efforts, their will, or their social circumstances, but because they are genetically programmed for their fate.

Evil in the Genes

The existence of evil has posed problems for philosophers and theologians for much of human history. Religious systems have personified evil as a supernatural being; folklore has located it in natural disaster, mythical beasts, or the "evil eye."[13] Evil can be seen as the cosmic consequence of fate (the [bad] "luck of the draw") or the result of voluntary human action or moral failure. The agents invoked to explain the presence of evil are commonly powerful, abstract, and invisible—demons, gods, witches, a marked soul, and, today, the biochemistry of the brain. Environmental contingencies, similarly powerful and abstract—such as patterns of authority (Stanley Milgram) or social reinforcement (B. F. Skinner)[14]—have also been seen as the sources of evil. But the belief that the "devil made me do it" does not significantly differ in its consequences from the belief that "my genes made me do it." Both seek to explain behavior that threatens the social contract; both locate control over human fate in powerful abstract entities capable of dictating human action in ways that mitigate moral responsibility and alleviate personal blame.

The response to research on the so-called criminal chromosome suggests the appeal of this view. In 1965 the British cytogeneticist Patricia Jacobs found that a disproportionate number of men in an Edinburgh correctional institution, instead of being XY (normal) males, were XYY males. Jacobs suggested that the extra Y chromosome "predisposes its carriers to unusually aggressive behavior."[15] Other researchers later questioned whether XYY males were more aggressive, suggesting instead that they suffered from diminished intellectual functioning that made it more likely that they would be incarcerated. And the original estimate of the rate of XYY males occurring in the population in general was later revised upward, so that the difference in the prison population and the general population appeared to be less great than it had once seemed.

But the "criminal chromosome" had a remarkable popular life, first attracting the attention of the press in April 1968 when it was invoked to explain one of the most gruesome crimes of the decade. A *New York Times* reporter wrote that Richard Speck, then awaiting sentencing in the murder, one night, of nine student nurses, planned to appeal his case on the grounds that he was XYY. This story— which was incorrect (Speck was an XY male)—provoked a public debate about the causes of criminal behavior. *Newsweek* asked if criminals were "Born bad?" ("Can a man be born a criminal?") *Time* headlined a story "Chromosomes and crime."[16] By the early 1970s, at least two films had featured an XYY male criminal, and a series of crime novels had made their focus an XYY hero who struggled with his compulsion to commit crimes.[17]

References to the criminal chromosome continued to shape popular views of violence. In 1986, the *New York Times* asked "Should such persons [XYY males] be held responsible for their crimes, or treated as victims of conditions for which they are not responsible, on a par with the criminally insane?"[18] And in 1992, a PBS series on "The Mind" introduced a segment on violence: "Recent research suggests that even the acts of a serial killer may have a biological or genetic basis."[19] Similarly, Donahue advised his listeners on "how to tell if your child is a serial killer." His guest, a psychiatrist, described a patient who had been raised in a "Norman Rockwell" setting but then, driven by his extra Y chromosome, killed 11 women.[20]

News reporters and talk show hosts refer to "bad seeds,"[21] "criminal genes,"[22] and "alcohol genes"[23]; CBS talk show host Oprah Winfrey found it meaningful to ask a twin on her show whether her sister's "being bad" was "in her blood?"[24] In the movie *JFK* one character tells another, "You're as crazy as your mama—Goes to show it's in the genes."[25] To a *New York Times* writer, evil is "embedded in the coils of chromosomes that our parents pass to us at conception."[26] And Camille Paglia described her theory of nature as following Sade rather than Rousseau: "Aggression and violence are primarily not learned but instinctual, nature's promptings, bursts of primitive energy from the animal realm that man has never left. . . . Dionysus, trivialized by Sixties polemicists, is not pleasure but pleasure-pain, the gross continuum of nature, the subordination of all living things to biological necessity."[27]

Genetic or biological explanations of "bad" behavior are sufficiently prevalent to serve as a common source of irony. A segment of the comic strip "Calvin and Hobbes" featured Calvin's perplexed father asking his son: "You've been hitting rocks in the house? What on earth would make you do something like that?" Calvin replied: "Poor genetic material."[28] In another strip, Calvin described a vicious "snow snake": "I suppose if I had two Y Chromosomes I'd feel hostile too!"[29] And a barroom cartoon by Nick Downes portrayed "Dead-Eye Dan, known far and wide for his fast gun, mean temper and extra Y chromosome."[30]

Bad genes have also become a facetious metaphor to describe national aggression. A *Time* article on the "new Germany" described the nation as "a child of doubtful lineage adopted as an infant into a loving family; the child has been good, obedient, and industrious, but friends and neighbors are worried that evil genes may still lurk beneath a well-mannered surface."[31] A journalist describing violence at English soccer games blamed it on the "genetic drive to wage war against the outlander."[32]

Some individuals, so the media imply, are "born to kill" and will do so despite environmental advantages. In December 1991 a fourteen-year-old high school boy was arrested for the murder of a schoolmate. The *New York Times* account of this event interpreted it as a key piece of evidence in "the debate over whether children misbehave because they had bad childhoods or because they are just bad seeds." The boy's parents had provided a good home environment, the reporter asserted; they had "taken the children to church almost every

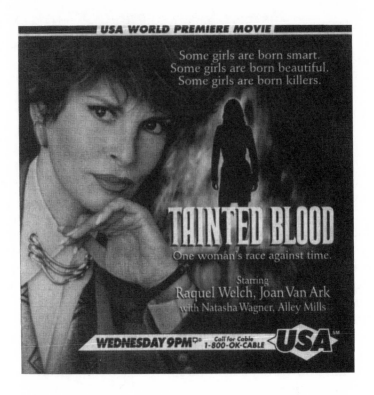

USA WORLD PREMIERE MOVIE

Some girls are born smart.
Some girls are born beautiful.
Some girls are born killers.

TAINTED BLOOD

One woman's race against time.

Starring
Raquel Welch, Joan Van Ark
with Natasha Wagner, Alley Mills

WEDNESDAY 9PM℗ᵉ Call for Cable 1-800-OK-CABLE USA℠

Sunday, and sacrificed to send them to a Catholic grammar school." Yet their son had been arrested for murder. This troubling inconsistency between the child's apparently decent background and his violent behavior called for explanation. The reporter resolved the mystery through the explanatory power of inheritance: the moral of the story was clearly stated in its headline: "Raising Children Right Isn't Always Enough." The implications? There are, indeed "bad seeds."[33]

A 1993 prime-time television movie called *Tainted Blood* drew on a related set of ironies.[34] A seventeen-year-old boy from a "stable" family killed his parents and then himself, shocking the community. The case attracted an investigative reporter (played by Raquel Welch) who found that the boy had been adopted and that his genetic mother had been in a mental institution. Suspecting he might have "inherited the gene for violence," the reporter went to the institution, where an elderly doctor (portrayed as ignorant of modern science) insisted that her ideas about heredity were wrong and that teenagers who murder were invariably abused. But it was the doctor who was wrong. The reporter discovered that the mother, like the son, had killed her parents and, eventually, herself. The boy had a twin sister who had also been adopted; thus began an urgent search for the girl with "tainted blood."

The scene changed to a suburb where Tory and Lissa, two seventeen-year-olds—both adopted—were best friends. Lissa was badly abused by her alcoholic mother, yet she was an honor student. Tory's adoptive family was stable, kind, and caring, but she was rather selfish and mean. When someone murdered Lissa's parents, Lissa, as an abused child, was immediately suspect. But it was Tory—genetic sister of the boy described above—who had murdered them in cold blood. Tory had also once threatened her younger adopted brother with scissors, but in a struggle between nature and nurture, nurture prevailed; she pulled back before she hurt him. Ultimately, however, "nature"—her predisposition to violence—won: after threatening her adoptive family again, Tory killed herself. The T.V. movie concluded when the reporter wrote a book on the incident and dedicated it to Tory; she was not to be blamed for her actions, because she had inherited a "genetic disease": she, like her twin brother, was a genetic victim. She had been "born to kill."

This same theme appeared in the news reports of a debate over the body and blood of Westley Allan Dodd, a serial killer of children who was hanged in January 1993. Dodd insisted that he could not be cured and that if he had the opportunity he would kill again and "enjoy it." His ordinary childhood offered no convincing explanation for his monstrous behavior. He had not been an abused child. After Dodd's execution, scientists attempted to obtain pieces of his brain and vials of his blood to determine whether his behavior could be attributed to neurological abnormalities or "gene oddities."[35] Such stories arise from a conflict between childhood experience and adult behavior; when the two seem to conflict, biological predisposition seems to provide a plausible and appealing resolution.

Research that links criminal behavior to biological forces fuels the hope that genetic information will make possible the prediction, and therefore the control, of deviant behavior.[36] Certain scientists encourage such expectations. In a 1992 *Science* editorial the journal's editor, biologist Daniel Koshland, told stories about acts of violence: "An elephant goes berserk at the circus, an elderly pillar of the community is discovered to be a child molester, a man admits to killing many young boys . . . a disgruntled employee shoots seven co-workers." Each crime, wrote Koshland, had a common origin—an abnormality of the brain.[37]

Some researchers who study aggression pique media interest by bringing their work directly to the public. In a *Psychology Today* article called "Crime in the Family Tree," behavioral psychologist Sarnoff Mednick reported that studies of adopted twins demonstrate the importance of heredity as a cause of crime: Among those boys with a biological parent who had a criminal record, 20 percent were themselves convicted. Where an adoptive parent had a criminal record, only 14.7 percent had been convicted.[38] In a *New York Times* essay called "The Aggressors," anthropologist Melvin Konner stated that the tendency for people to do harm to others is "intrinsic, fundamental, natural."[39] Such reports from scientific authorities imply there is definitive evidence for the importance of genetic predisposition as a cause of criminal behavior.

Studying the media coverage of XYY research, sociologist Jeremy Green has described how scientists fostered images of the "XYY Man" through the expository strategies they employed to publicize their work.[40] The press disseminated these images through provocative stories, perpetuating belief in "criminal tendencies" and hope of prediction and control. For example, in response to a reader's inquiry about the possibility of finding out if someone has a criminal tendency, medical columnist Lester L. Coleman wrote (incorrectly) in 1990: "The genetic code of life has been clearly established by scientists who recently were awarded the Nobel Prize for this brilliant achievement. A part of their study was devoted to the abnormality of an extra X or Y chromosome. The value of this important knowledge may yet lead to the possibility of predicting criminal tendencies."[41]

Even when scientists emphasize the complexity of biological and environmental conditions that could lead to violence, media accounts highlight the importance of genetics. The press coverage of the National Research Council's 1992 report on the state of research on violence is a case in point.[42] The report said that violence arises from the "interactions among individuals' psychosocial development, neurological and hormonal differences, and social processes." It stressed the uncertain implications of research when it came to genetic influence on antisocial behavior: "These studies suggest at most a weak role for genetic processes in influencing potentials for violent behavior. The correlations and concordances of behavior in two of the three studies are consistent with a positive genetic effect, but are statistically insignificant." While not ruling out genetic processes, the NRC suggested: "If genetic predispositions to violence are discovered, they are likely to involve many genes and substantial environmental interaction rather than any simple genetic marker."

Only 14 of the 464 pages of the NRC report actually dealt with biological perspectives on violence, and less than two pages were about genetics. Nevertheless, the *New York Times* headlined its article about the report "Study Cites Role of Biological and Genetic Factors in Violence."[43] Genes appear far more newsworthy than social or economic circumstances as a source of anti-social behavior. While genetic theories of violence have been controversial, denounced as politically and racially motivated (see Chapter 6), some journalists have dismissed critiques as "politically correct." In April 1993 *Time*, looking for the causes of "the savagery that is sweeping America," suggested that society's ills cannot fully be responsible, that violence may be caused by "errant genes." "Science could help shed light on the roots of violence and offer new solutions for society. But not if the research is suppressed."[44]

Biological theories also appeal as explanations of group violence and war. A 1991 social psychology textbook uses "genetic similarity theory" to explain "the tendency to dislike members of groups other than our own." Discrimination against those who are different, say the authors, is part of inherited human tendencies to defend those possessing similar genes.[45] Extending this idea to explain war, a *Discover* journalist described a study of chimps and speculated whether "war runs in our genes like baldness or diabetes?"[46] Such explanations extend the popular theories of the late 1960s and early 1970s, when a spate of books appeared explaining human behavior to a lay audience in evolutionary terms. These included Robert Ardrey's *The Territorial Imperative* (1966), Konrad Lorenz's *On Aggression* (1966), Desmond Morris's *The Naked Ape* (1967), and Lionel Tiger and Robin Fox's *The Imperial Animal* (1970). Promoting a biological model of organized human aggression, these authors explained it as a productive and necessary social activity.[47] The books were fashionable, attracting a wide readership and extensive media coverage. Reviewing the response to aggression research, Temple University psychologist Jeffrey Goldstein found that the media systematically covered studies that offer evidence of genetic explanations of violence, but were less interested in research on the influence of social and economic conditions.[48]

Some biologists and social scientists have criticized research on the genetic predisposition to organized aggression for concealing inadequate methodologies behind quantitative data and for minimizing the influence of social, political, and economic factors on aggressive behavior. In May 1986, Goldstein helped assemble a group of these critics to discuss biological theories about the origin of warfare. Meeting in Spain, they produced the Seville Statement on Violence, which strongly repudiated the idea that war is biologically necessary or genetically controlled. "It is scientifically incorrect to say that war is caused by 'instinct' or any single motivation . . . scientifically incorrect to say that humans have a 'violent brain' . . . scientifically incorrect to say that in the course of human evolution there has been a selection for aggressive behavior more than for other kinds of behavior." The statement concluded that "biology does not condemn humanity to war. The same species who invented war is capable of inventing peace."[49]

This brief but unambiguous text was signed by twenty well-known scholars from around the world and endorsed by the American Psychological Association, the American Anthropological Association, the International Society for Research on Aggression, and Psychologists for Social Responsibility. Yet despite considerable efforts to publicize the statement, it attracted little media attention. A journalist responding to the efforts to disseminate the Seville material expressed the prevailing attitude: "Call me when you find the gene for War."[50]

The interest in "bad genes"—the genes for deviance— reflects a tendency to medicalize social problems.[51] This is especially evident in scientific and social speculation about the nature and etiology of addiction. Definitions of alcoholism have shifted over time from sin to sickness, from moral transgression to medical disease, depending on prevailing social, political and moral agendas.[52] Debates over the etiology of alcoholism go back to ancient Rome, but the modern conception of alcoholism as a disease is usually attributed to the nineteenth-century theories of Benjamin Rush (1745–1813). Early leaders of the American temperance movement, likewise, defined alcoholism as a disease, but when the movement began to advocate outright prohibition, alcoholism was redefined, along with syphilis and opiate addiction, as a "vice"—a manifestation of immoral behavior. A moral concept of voluntary addiction replaced the model of disease, and the politics of prohibition in the 1920s turned alcoholism into a problem more legal than medical. At the same time, eugenicists were compiling family studies supposedly demonstrating its inherited nature.

In 1935 E. M. Jellinek, reviewing the biological literature on alcoholism for a major Carnegie Foundation report, formulated a medical model that explained alcoholism in terms of the interaction of alcohol with an individual's physical and psychological characteristics and his or her social circumstances.[53] This analysis focused attention on what made people susceptible. The same year Alcoholics Anonymous was founded, on the doctrine that alcoholism was a compelling biological drive that could be cured only by total abstinence and moral rectitude. AA's position contributed to the revival of the medical model, promoting the idea that alcoholics had "predisposing characteristics" that distinguished them from others.[54] This view has persisted, in its contemporary form focusing on the genetic basis of alcoholism.

Common observation shows that alcoholism runs in families. As in the case of violence, however, this in itself does not reveal the cause. Many traits run in families—poverty, for example, or poor manners—without being a consequence of heredity. The prevalence of alcoholism in certain families could reflect role models, the availability of alcohol, or the reaction to abuse. Nevertheless, a common perception was expressed in a 1989 article in *Omni*: "Addicted to the bottle? It may be in your genes."[55] The gene for alcoholism became a theme on the Oprah Winfrey and Phil Donahue shows. A 1990 article in *Mademoiselle* asked: "Do you have a gene that makes you a designated drinker?" and suggested that "even if you have exceptional self-discipline, you could still be at high risk."[56] And Nancy Reagan's famous anti-drug slogan, "Just say No" provoked a 1991 editorial about the "genes-impelled compulsion" to take drugs.[57]

In a 1990 article on addiction, *New York Times* reporter Daniel Goleman presented several cases to dramatize the genetic basis of alcoholism. A 26-year-old executive had been the class clown as a child and president of his high school class. Always extroverted and outgoing, he partied a lot and, as he matured, started taking drugs in order to stay high. Addiction appealed because of his "natural bent." Another young man had been anxious as a child until he discovered that alcohol made him relax. His father was an alcoholic, so he had easy access to liquor; Goleman, however, quoted sources that explained his addiction in terms of biological vulnerability.[58]

Goleman's stories suggested one reason for the appeal of genetic explanations: they implied that biological markers will make it possible to identify those at risk of addiction. He quoted a scientist who optimistically claimed that genetic engineering will eventually eliminate the gene and therefore the problems of addiction. In effect, like genetic explanations of violence, identifying an "alcoholism gene" offers the hope that addiction can be controlled—not through the uncertain route of social reform, but through biological manipulation.

Assumptions about the genetic basis of alcoholism have extended to other addictions: smoking, overeating, shopping, and gambling. Some news reports on diet control begin from the assumption that genetics is the underlying reality that determines obesity; "Where Fat is a Problem, Heredity is the Answer," read a 1990 headline in the *New York Times*.[59] In another *Times* story, this one on smoking, a journalist wrote: "Smoking has to do with genetics, and the degree to which we are all prisoners of our genes. . . . You're destined to be trapped by certain aspects of your personality. The best you can do is put a leash on them."[60] And in a "Dear Abby" column, a lifelong smoker announced that he had no intention of stopping. His reasoning? "Heredity plays a major role in how long we live—not diet and exercise, jogging and aerobics, or any of the other foolishness that health freaks advocate." His father was 86 and in perfect heath, so he felt free to smoke, "eat ham and eggs fried in butter," and "steak and baked potatoes with plenty of sour cream."[61]

Self-help books devoted to coping with addiction speculate on why some people are affected by their circumstances more than others. The answer: biological predisposition. Addicts (those who shop, eat, love, or drink too much) are "victims of a disease process," and the disease—the tendency towards compulsive behavior—is transmitted by their families, says one expert. We read that "Obsessive Compulsive Disorder is partially genetically transmitted. . . . Most researchers agree OCD will develop only if an individual is genetically predisposed to it."[62] Similarly, in a guide written for the families of gamblers, we learn that some people have a "biological temperament" that makes them especially susceptible to addictions.[63]

To explain addictive behavior in absolute genetic or biological terms is to extract it from the social setting that defines and interprets behavior. There are no criminal genes or alcohol genes, only genes for the proteins that influence hormonal and physiological processes. And only the most general outline of social behavior can be genetically coded.[64] Even behaviors known to be genetically inscribed, such as the human ability to learn spoken language, do not appear if the environment does not promote them. Children do not learn to speak unless they hear spoken language, even though the ability to speak is genetic, a biological trait of the human species. In the case of alcoholism, for example, any biological or genetic predisposition that may exist can only become a full-blown pattern of behavior in an environment in which alcohol is readily available and socially approved. As this suggests, there are many interests at stake in the etiology of addiction, for causal explanations for addiction imply moral judgments about responsibility and blame.

If defined as a sin, alcoholism represents an individual's flaunting of social norms; if defined as a social problem, it represents a failure of the community environment; if defined as intrinsic to the product consumed, it represents the need for alcohol regulation. But if defined as a genetically determined trait, neither society nor the alcohol industry appears responsible. And if behavior is completely determined—either by genetics or environment—even the addicted individual cannot really be blamed.

Just as addiction, crime, and war appear to be molecular events writ large, so too are special talents, the remarkable success of celebrities, and even the qualities of inanimate objects such as automobiles, perfumes, or magazines.

Good Genes

Sixteen-year-old Judit Polgar is the youngest chess grand master ever, and the first female chess player considered to have the potential to be a world champion. Her parents insist that her talent is a consequence of family training. She and her two sisters, all world-class chess players, have never attended school but are tutored at home by their multilingual mother and their psychologist father, who believes that "every child is a potential genius." He raised his children as an "experiment" in the power of environment, and it seems to be a successful experiment: Judit and her sister Zsuzsa are grand masters, ranked respectively as the first and second female chess players in the world; their middle sister, Sofia, is ranked sixth. But according to a journalist's 1992 account of the remarkable Polgar sisters, their father's environmental explanation is "looked upon with skepticism" in the chess world. Another grand master is quoted as saying: "I think they were born gifted, and one is a genius."[65]

There are many possible explanations for success; hard work, persistence, talent, the exposure to role models, the availability of contacts and professional opportunities, the social pressures from family or peers, or simply good luck. Ubiquitous American narratives of success draw on the Horatio Alger myth to suggest that any person who puts in the effort can "make it." The myth of the "Jewish mother" constructs guilt in the Jewish family as a driving force in the success of children. And the recent success of Asian students has evoked stories about parental and ethnic community pressures stressing hard work and achievement.

But another set of narratives suggest that achievements have genetic origins, that success is biologically determined. Just as children from "good" families may turn out "bad," so those with limited opportunity can, with the proper genes, rise above their circumstances. A television newscaster in 1988 described a teenager named Mike who, though raised in a poor family with no father, became captain of his track team and won a college scholarship: "He has a quality of strength and I guess it has a genetic basis."[66] A 1991 *Newsweek* article explained how poverty, physical impairment, or abuse affects children: "Some kids have "protective factors that serve as buffers against the risks." They have "natural resilience" or "built-in defenses." It is the "genetic luck of the draw."[67]

Genetic explanations of socially valued traits also frequently appear in stories about popular personalities, whether scientists, actors, sports heroes, politicians, or rock musicians. The gene is appropriated in accounts of famous parent-child pairs, in stories of famous scientists, and in the voyeuristic reporting about the spectacular ascent of media personalities. We read about Elvis's genes and Einstein's DNA.

In a culture obsessed with fame, money, and personal success, the cause of exceptional achievement is a matter of widespread curiosity. Why are some people more successful than others? What accounts for extraordinary achievements? Genetic explanations of success reflect widely held beliefs about the importance of heredity in shaping special talents. Such explanations imply a change in the "bootstrap ideology" that has pervaded American folklore, for they undermine the myth that an individual's will or hard work alone determines his or her achievement or success. Neither individual actions nor social opportunity really matter if our fate lies in our genes.

The range of special talents attributed to genetics is remarkable. In a story about the Ginsberg brothers, both of whom are poets, the *New York Times* referred to their "poetry genes."[68] An obituary writer explained the secret of Isaac Asimov's success: "It's all in the Genes."[69] A story about a promoter of rap music described his style: "Charlie Stettler has no embarrassment gene."[70] Body-building books insist on the "natural genetics of muscular men. The mesomorphs are chosen people" with "superior genetics."[71] A 1992 fashion headline asked, "Is Good Taste in the Genes?"[72] Former New York City comptroller Elizabeth Holtzman explained why she was attracted to social causes: "It's in my genes."[73]

Genetic explanations appear in unlikely places. A 1993 biography of James Joyce contained a diagram of Joyce's "genetic makeup."[74] A horoscope columnist announced that she was "genetically coded" with impeccable social instincts.[75] A gardener celebrated her "gardening genes."[76] Mother's Day cards use quips about genetics: my imperfections "must be the genes from Dad's side of the family." Or "Congratulations to a daughter who turned out to be a fine mother. Aren't Genes Wonderful?" A 1993 Maybelline cosmetics commercial featured the fashion model Christy Turlington while two questions flashed on the screen: "Is it in her genes?" and "Was she born with it?"

The metaphorical power of the gene is especially striking when inanimate objects—particularly automobiles—are marked as "successful" by virtue of genes. A Sterling's remarkable handling is "in its genes;"[77] a BMW sedan has "a genetic advantage;"[78] a Subaru is a "genetic superstar;" and a Toyota has "a great set of genes."[79] The Infinity has DNA that defines its authenticity: "While some luxury sedans just look like their elders, ours have the same DNA."[80]

Genes mark the quality of other products as well. A Nike sneaker "has inherited its own set of strength . . . resilience . . . stability, and a true intelligent fit."[81] A Bijan perfume called "DNA" is advertised as "a family value," inspired by the "power of heredity." It is "the stuff of life."[82] A blue jeans ad exploited the obvious pun ("Thanks for the genes, Dad"), while implying their superior quality.[83] An article on the leadership changes in *The New Yorker* asked: "Can you change a magazine's DNA? . . . A magazine's underlying character remains—unchanged and enduring, a DNA-like set of fingerprints—and lasts through the years and reinventions. . . . Tina Brown has much to reckon with, starting with 67 years of DNA."[84]

If good genes appear in popular culture to describe the quality and value of consumer products, they can also turn people—in particular, children—into products as well. Magazine stories entitled "The Baby Shoppers," "Babies for Order," or "Looking for Mr. Good Genes" portray the infant as a commodity. One writer described a sperm bank as "a specialty shop . . . for people who want to do boutique shopping."[85] The search for the right sperm in these stories becomes a kind of catalog shopping; potential parents scan lists of desirable traits. Sperm donor profiles read like personal ads, providing detailed information about the donor's favorite colors, ability to carry a tune, and hobbies. Sperm bank sagas deal with reproduction as an abstract and commercial transaction. Women are hosts or breeders, responsible for creating good stock. Sperm donors, as one writer put it, become "prize bulls" or "nuclear age studs."[86]

Shopping for good genes has also become a soap opera plot device. In 1991 and 1992 on "The Young and the Restless," Leanna wanted to have a baby with certain desirable characteristics, so she searched for "father prospects" who would donate their sperm.[87] In "Santa Barbara," Gina applied for a job at a sperm bank, because "What better way to find the daddy of her dreams?"[88] And in a 1992 episode of the television program "Doogie Houser, M.D.," a female doctor pursued the seventeen-year-old genius not because she was interested in him sexually, but because she "wanted" his genes.[89]

This depiction of children as products of their genes continues as they develop interests and careers. The media are attracted to celebrity families with similar professional interests. Children do often enter the professions of their parents, and they are attracted to these professions for many reasons, including social expectations, family pressures, the availability of unique opportunities, and personal contacts. But many accounts of children who follow the career paths of their parents emphasize their genetic heritage. As in the earlier narratives of the eugenics movement, a trait that is shared by both parent and child is assumed to be genetic. For example, a *Rolling Stone* review of two jazz saxophonists, Josh Redman and Ravi Coltrane, focused on their similarities to their fathers, both of whom had been sax players. The social influences on the boys' career choices were clear; they had listened to jazz and developed strong family contacts who, the article noted, had helped them get jobs. But the review was about their DNA: "Maybe it's in the genes."[90]

When Ringo Starr, former drummer for the Beatles, was interviewed on "The Arsenio Hall Show," the host commented on the interests of Mr. Starr's son. "He's a drummer too? He must have the drummer gene."[91] A host of "The Today Show" introduced the daughter of singer Marvin Gaye as having her father's "talent genes."[92] And in the CBS coverage of the 1992 Olympic ice skating competition, the announcer said Japanese skater Yuko Sato had a "genetic advantage" because her parents were skaters too.

Even William Safire, among the most self-conscious users of language, has been affected by the popular fascination with genes for success. In a 1992 *New York Times* column about the romance between Albert Einstein and Miliva Maric, Safire speculated about what happened to their daughter, who had been given up for adoption: "We can presume she grew up to have a family of her own and that humanity has been enriched by the propagation of the genes of genius."[94]

Safire's comment reflects the considerable professional and popular curiosity about the origins of genius. Just as Dodd's brain was coveted as a way to understand his criminal fate, so Einstein's brain has been preserved and the fragments studied to discover the biological peculiarities of his genius. In 1993, *Skeptic* magazine published a special section on genius, asking people they felt had special insights to share their thoughts. Comedian Steve Allen was confident that science will show that genius is genetic. Marilyn Vos Savant, listed in the *Guinness Book of World Records* as the person with the highest IQ, said that "genius must begin in the genes." Biochemist Elie Shneour said in 1993 that she expected genius to be mapped in the human genome.[95]

Genes are also commonly used to characterize the foibles of successful politicians. Journalists described former President George Bush as missing an empathy gene[96] and Presidential candidate Ross Perot as possessing a "frugal gene" (Perot had insisted that his children take their own popcorn to the movies rather than wasting their allowance on overpriced theater concessions).[97] When Pat Buchanan was running as a Presidential candidate, a reporter referred to one of his aides as a "genetic conservative."[98] Before the 1992 election, a political joke suggested that Democratic men in Washington were dating Republican women in order to replenish their gene pool so that they could produce a winner.[99] Describing the role of the wives of Presidential candidates, columnist Anna Quindlen defined the "missing gene theory of political marriages: She must provide something he lacks." Thus, she suggested, Nancy Reagan carried Ronnie's retribution gene; Mrs. Bush carried George's compassion gene. "Mrs. Quayle has more to do; it's said she carries the brain for the couple. The idea is that spliced together the husband and wife form a much more perfect union."[100]

The gene has also been appropriated to explain the personalities of those in business. A business writer described successful entrepreneurs as having inherited business tendencies.[101] And in an article on the dispute over the ownership of the *New York Post,* a journalist characterized the controversial candidate Abraham Hirschfeld as lacking "the gene for self-censorship, as he is constantly saying—or spitting—the wrong thing at the wrong time."[102]

The media often suggest that successful people pay for their genetic legacy: genes are elements in a tragedy, since human greatness inevitably has a price. As in the story of Elvis, writers draw an association between genius and mental illness or alcoholism.[103] Stories about actress Drew Barrymore, the daughter of John Barrymore Jr., for example, blamed her problems on her biological heritage. Drew, a former alcoholic, was described in the press as "the latest inheritor of her family's dark legacy of alcoholism and addiction."[104] Her mother had left John Barrymore, Jr., when she was pregnant; Drew, according to the media, had an unhappy family life. She remembered her father hurling her against a wall and putting her hand through candle flames. But in the stories about her addiction, such traumatic childhood experiences were not the cause of her drinking—rather, genetic alcoholism blighted the thespian family tree. Drew herself described her problem: "I believe it is genetic. I was somehow destined."[105]

The appropriation of DNA—the good or bad gene—to explain individual differences recasts common beliefs about the importance of heredity in powerful scientific terms. Science becomes a way to empower prevailing beliefs, justifying existing social categories and expectations as based on natural forces. The great, the famous, the rich and successful, are what they are because of their genes. So too, the deviant and the dysfunctional are genetically fated. Opportunity is less important than predisposition. Some are destined for success, others for problems or, at least, a lesser fate. The star—or the criminal—is not made but born.

This is a particularly striking theme in American society, where the very foundation of the democratic experiment was the belief in the improvability, indeed, the perfectibility of all human beings. Belief in genetic destiny implies there are natural limits constraining the possibilities for both individuals and for social groups. Humankind is not perfectible, because the species' flaws and failings are inscribed in an unchangeable text—the DNA—that will persist in creating murderers, addicts, the insane, and the incompetent, even under the most ideal social circumstances. In popular stories, children raised in ideal homes become murderers and children raised in difficult home situations become well-adjusted high achievers. The moral? No possible social system, no ideal nurturing plan can prevent the violent acts that seem to threaten the social fabric of contemporary American life. Only biological controls, it seems, can solve such problems.

The idea of genetic predisposition encourages a passive attitude toward social injustice, an apathy about continuing social problems, and a reason to preserve the status quo. Genetic explanations, however, are malleable. They can be appropriated to justify prevailing stereotypes and maintain current social arrangements, but they can also be used to promote group identity or to celebrate human differences. The diverse social, political, and moral dimensions of such explanations become most transparent as they appear with growing frequency in stories and debates about the social meaning of sex, race, and sexual orientation.

6

Creating
Natural
Distinctions

Biological explanations have long been used to justify
social inequalities by casting the differential treatment and
status of particular groups as a natural consequence of
essential, immutable traits. In the 1990s the language of
genetic essentialism has given new legitimacy to such explanations. Group differences are appearing in popular culture
as genetically driven, encouraging stereotyped images of the
nurturing female, the violent African American male, and
the promiscuous homosexual. But the images of pathology
have moved from gross to hidden body systems. Once blacks
were portrayed with large genitalia and women with small
brains: today the differences lie in their genes.

The belief in essential differences has been reinforced by scientific studies of body parts such as genes or neurons that seem to explain behavior, as well as by scientific theories about evolution that seem to biologically ground social practices. Molecular genetics, behavioral genetics, neuro-biology, and sociobiology have provided a language through which group differences can be interpreted as biologically determined. These sciences have encouraged the increasing acceptability of genetic explanations and their strategic role in the continuing debates over gender, race, and sexual orientation.

Current interest in the genetic basis of group differences coincides with extraordinary concern about gender roles, ethnic identity, and sexual orientation. Genetic explanations can be used to marginalize groups or—as in the case of some feminists, African Americans, and homosexuals—to cele-brate group differences. Some who have traditionally suf-fered from prevailing biological theories are now embracing biological difference as a source of legitimacy and as evi-dence of their own superiority.

They shrewdly exploit the discursive power of biological boundaries to promote reformist agendas. Some feminists have celebrated biological difference as a source of special identity or a rationale for equal protection, citing the "cre-ative power that is associated with female biology" and the "native talent and superiority of women."[1] Controversial Afrocentrist Leonard Jeffries, a professor at City College in New York, has claimed that melanin is "responsible for brain development, the neurosystem and the spinal col-umn"; since African Americans have more of it, they are more creative.[2] Meanwhile gay activist Simon LeVay has promoted the idea that homosexuality is inborn and un-changeable, for such a claim seems to transform non-conformist sexual behavior from a "lifestyle choice" to a natural imperative.[3] These individuals, despite radically dif-ferent perspectives and conflicting social policy agendas, seem to agree about one thing: In contests over social worth, biology matters. Whoever can successfully argue that biology—and more specifically DNA—supports their particular political viewpoint has a tactical advantage in the public debate.

Genes and Gender

The January 20, 1992, cover of *Time* featured a photograph of two small children, a girl and a boy. The girl looked admiringly and flirtatiously at the boy, who was flexing his muscles. The caption read: "Why are men and women different? It isn't just upbringing. New studies show they are born that way." The cover story, called "Sizing up the Sexes," reported that recent scientific research showed that "gender differences have as much to do with the biology of the brain as with the way we are raised."[4] The article focused on an experiment in which the responses of young children playing with a variety of toys were videotaped. The researcher found that the boys systematically favored sports cars and fire trucks, while the girls were drawn to kitchen toys and dolls. Those girls who defied these expectations, researchers concluded, had higher levels of testosterone, suggesting the importance of hormone levels in shaping the behavior of children.

Time went on to report a general public conversion to the idea that differences between men and women are genetically determined. Even skeptics in the professions, the article said, had come to see that nature dictated social place and behavior. It quoted psychologist Jerre Levy to illustrate a typical intellectual shift: "When I was younger, I believed that 100% of sex differences were due to the environment." But now, observing her fifteen-month-old daughter, she believed that skill in flirting must be inborn. The *Time* writer editorialized: "During the feminist revolution of the 1970s, talk of inborn differences in the behavior of men and women was distinctly unfashionable, even taboo. Men dominated fields like architecture and engineering, it was argued, because of social, not hormonal, pressures. Women did the vast majority of society's childrearing because few other options were available to them." Feminists, said the author, expect that the "end of sexism" will change all this: "But biology has a funny way of confounding expectations. . . . Perhaps nature is more important than nurture after all."[5]

The 1992 *Time* story differed in striking ways from a *Newsweek* cover story on the same subject a decade earlier. Its title ("Just How the Sexes Differ"[6]) was similar, and like the story in *Time*, it attempted to convey prevailing scientific wisdom about the relative role of nature and nurture in shaping differences in the behavior of men and women. It even drew on some of the same authorities and referred to some of the same scientific data. The 1981 article, however, stressed the overriding importance of different social experiences and expectations on behavior. The author cited several experts on the effect of culture on behavior, including anthropologist Sarah Blaffer Hrdy, who said gender differences were the product of culture, and Michael Lewis of the Institute for the Study of Exceptional Children, who said: "As early as you can show me a sex difference, I can show you culture at work." And it cited psychologist Eleanor Maccoby: "Sex typing and the different set of expectations that society thrusts on men and women have far more to do with any differences that exist . . . than do genes or blood chemistry."

The *Newsweek* journalist pointed to the broad variation among individuals of each sex: "Not all males in a given group are more aggressive or better at math than all females. . . . women and men both fall along the whole continuum of test results." This 1981 article concluded with a statement on the nature-nurture debate: "It is clear that sex differences are not set in stone. . . . By processes still not understood, biology seems susceptible to social stimuli." Ten years later, *Time* conveyed the opposite message: sex differences are innate and unchangeable, and it is just a matter of time until scientists will be able to prove it.

These two approaches to the same subject matter, over such a short period, suggest shifts in the social meaning of science. Neither popular article was about scientific data per se. Rather, both captured popular expectations. By 1992 the consensus that had dominated public policy since the 1950s—that "nurture" was more important than "nature"—was changing.

Neither biological nor environmental explanations of human behavior have an inherent social meaning. Both forms of explanation can be used to justify liberal or conservative causes; both can be applied oppressively, and each can be used to promote greater human freedom. In the last two decades biological determinism has been the focus of several well-publicized attacks by leading academic biologists and philosophers. But environmental determinism, too, can be used to limit human rights and constrict groups identified as inferior. In the 1950s and 1960s, for instance, popular interest in the power of the environment reinforced women's traditional roles as caregivers. It justified the 1950s "back to the home" movement for mothers who had been employed during the war years; if the achievements of children were finely calibrated to their training and environment, then mothers were needed at home and entirely responsible for their children's behavior.

So too, evolutionary stories serve different social agendas. Some feminists have presented evolutionary narratives in an attempt to demonstrate the superiority of women, while sociobiologists have interpreted evolutionary history as legitimating inequalities. Elaine Morgan's 1972 account of "sexual evolution" was a popular and colorful story in which *Homo sapiens* spent a few million years as a marine mammal, during which the species lost its body hair, developed a layer of adipose fat, and invented tools and weapons. Women were important players in the development of these new technologies, Morgan said, for they needed tools to fish in the surf (babies clinging to the long hair on their heads) and to pry open shellfish. Morgan used this imagined evolutionary past to legitimate the sexual revolution of the 1960s and its promise for the "liberation" of women.[7]

Three years later entomologist and sociobiologist E. O. Wilson reached different conclusions about the meaning of the species' evolutionary past for current social stratification.[8] Drawing on his extensive work with social insects, Wilson extrapolated from the ant to the human to suggest that women will naturally seek monogamous unions and that they are biologically best suited to childrearing. Women are responsible for child care and home life in most societies because, Wilson argued, "the genetic bias is intense enough to cause a substantial division of labor even in the most free and most egalitarian of future societies."

Richard Dawkins, in *The Selfish Gene* (1976), took Wilson's arguments a step further to suggest that male priorities in the "battle of the sexes" would naturally lead to male promiscuity and fathers' abandonment of children. Women are responsible for child care, not because they are uniquely suited for the task, but because of the abstract forces of evolution.[9] He noted that mothers have a disproportionate biological stake in their newborns because they contribute more biological material (an egg) and serve as receptable for the growing embryo. Mothers and fathers, he said, may well have an equal genetic stake in their children, but at the moment of a baby's birth, their time investment differs dramatically. Women will therefore naturally—and wisely— be more interested in infants than are men. Dawkins's conclusions were welcomed by leaders of the new right. Phyllis Schlafly, for example, adopted genetic essentialism to support her opposition to the Equal Rights Amendment. Social inequalities, she said, necessarily follow fundamental biological differences between the sexes; women, after all, have babies and men must support them.[10] And in Minnesota, Allen Guist, 1994 gubernatorial candidate from the religious right, contended, "There is a genetic predisposition for men to be heads of households."[11]

Biological arguments have also been used to suggest that women are unsuited for certain kinds of work. In 1980, psychologists Camille Benbow and Julian Stanley published a technical research paper in *Science* on the differences between adolescent boys and girls in mathematical reasoning.[12] The *New York Times* cited the authors as urging educators to "accept the possibility that something more than social factors may be responsible. . . . You can't brush the differences under the rug and ignore them."[13] The research inspired so many popular articles that in 1980 the *Reader's Guide* added a new heading to its listings: "Math ability."

"Because my genetic programming prevents me from stopping to ask directions—that's why!"

Time announced that the "gender factor in math" meant that "males might be naturally abler than females."[14] *Discover* reported that male superiority at mathematical reasoning was so pronounced that "to some extent, it must be inborn."[15]

Genetic arguments, however, can be strategically deployed to justify quite different visions of appropriate gender roles. Some narratives glorify women for precisely the same reasons that other narratives assign them lower status: their biological uniqueness. At the same time, they transform the male qualities associated with testosterone—aggression and domination—into socially negative traits, interpreting them as inherently, biologically, male. One feminist group in 1973 argued that men dominated women because "men are innately more aggressive than women, due to the effects on the brain of the male hormone testosterone." As Alison Jagger has noted, this was precisely the prevailing explanation of male supremacy that appeared in Steven Goldberg's antifeminist text *The Inevitability of Patriarchy* the very same year.[16]

In the 1980s and 1990s, it has become increasingly fashionable to presume the existence of innate and intrinsic differences between the sexes. Even books in the self-help genre (by definition less deterministic, concerned as they are with issues of personal control) use biological ideas to enhance the confidence of their women readers and to instill group pride. Elizabeth Davis, in *Women's Intuition* (1989), proclaimed that "intuition is a natural function, and that all of us [women] have inherent intuitive abilities. . . . Women are more sensitive to context, more receptive."[17] And in their 1990 popular advice book *Why Women Worry*, Jane and Robert Handly cited scientific findings about differences in brain size that they claim confirm women's natural role as protectors of children and men's role as hunters.[18]

Purported differences in the brain were also the focus of Ann Moir and David Jessel's 1991 *Brain Sex: The Real Differences Between Men and Women*, a popular book and television program. Citing research on nonhuman primates, its authors claimed that society should accept biological differences and build them into early childhood education.[19] An article in *Mirabella* magazine, describing this book, featured a photograph of a nude woman. The writer asked: "Are women genetically programmed to enjoy pleasure? A new book . . . argues, yes, . . . Women hear better, are more tactile, even see better in the dark."[20] The women's magazine *Elle*, describing the book in an article called "Mind Over Gender," concluded that: "It just isn't the case that child rearing practices are the only influence on how we grow up. . . . The primary determinant of what and who we will turn out to be is to be found in our bodies. In other words, the old nature-versus-nurture quibble has finally been settled—and nature now appears to be the winner."[21]

Child rearing advice books have picked up this theme. Sheila Moore and Roon Frost, in their 1986 *Little Boy Book*, told parents "Yes, boys are different with their built-in combativeness and delayed impulse control." Citing the work of developmental psychologist Sandra Scarr, the authors, both mothers of boys, stated that boys are genetically programmed for aggressiveness, independence, "horse play," and learning disabilities.[22]

The naturally aggressive male also appears in the debate over gender and war. As women's military roles have expanded, the absence of "war" genes in women has been cited as a reason to limit their roles in the American military hierarchy. In 1986, neoconservative economist George Gilder wrote that hard evidence showing that men are more aggressive means that women should not go into combat. Gilder suggested that permitting women to participate in combat would put the United States in mortal danger, since the USSR not only had a larger army, it had one that was entirely male.[23] Similarly, a 1990 letter to the *New York Times* used sociobiological arguments to oppose the drafting of women. The male is "hard wired through evolution to defend the tribe. . . . A woman can have 10 to 15 surviving babies in her reproductive lifetime," so survival of societies has depended on the fact that males rather than females are predestined biologically to go to war.[24] A political consultant during the United Nations "Year of the Woman" (1992) proclaimed that "women are by nature less inclined to throw the first punch."[25] And the cultural critic Camille Paglia framed the same idea in different terms: "I can declare that what is female in me comes from nature and not from nurture. . . . The traditional association of assertion and action with masculinity, and receptivity and passivity with femininity seems to me scientifically justified. . . . Man is contoured for invasion, while woman remains the hidden, a cave of archaic darkness."[26]

In 1991 Robert Bly's *Iron John* became the top nonfiction best-seller and the inspiration for a new television sitcom.[27] Bly wrote about men and women as virtually different biological species, suggesting that men and their natural characteristics have been weakened and feminized by culture. In language reminiscent of the Nazi youth movement, he found manhood in rituals of the body. His book gave rise to a social movement that focused on "real" men and—in almost Nietzschean terms—the "beast" within the man. Movement participants have organized wilderness retreats where men "enter some of the boggy depths of maleness in search of the wild man deep inside." A *New York Times* review of an ABC sitcom about the movement was headlined: "In Touch with the Tool Belt Chromosome."[28]

This celebration of natural male aggression and "beast-liness" has a feminine corollary. Clarissa Pinkola Estes's 1992 *Women Who Run with the Wolves: Myths and Stories of the Wild Woman Archetype* presented a "psychology of women in the truest sense," intended to help her readers connect to the "innate, basic nature of women." This nature, she said, has been suppressed by society, which has "muffled the deep, life-giving messages of our own souls." Echoing Bly's comments about the effects of society on men, she argued that women have been "over-domesticated" by culture so that they lose touch with their natural, fundamental selves. She recommended story-telling to help women reconnect to the "healthy, instinctual, visionary attributes of the Wild Woman archetype."[29] Her work has, like Bly's, inspired a series of retreats at which participants are encouraged to connect with nature. At one of these "Wild Woman" wilderness retreats, an organizer announced that women are inherently "aesthetic, still, intuitive, nonverbal, nurturing," and not prone to "rational thinking." Her seminar was intended to help women bond.[30]

The message that differences between men and women can be localized to genes sometimes appears in less direct ways. In scientific textbooks, popular articles, and Hollywood movies, the germ cells of reproduction behave suspiciously like lovers in a romantic folk tale. The sperm appears as active and competitive, seeking out the passive, coy egg, which it can reach only by conquering the "hostile environment" of the vagina and uterus. In a 1986 *National Geographic* article, sperm were described as "masters of subversion" employing "several strategies to survive their mission." In a *Far Side* cartoon, the egg appeared as a housewife besieged by clever sperm masquerading as a phone repairman, insurance salesman, and postman, all trying to get a foot in the door.[31] And *Discover*, reporting on studies of sperm motility, called "competitive sperm" the "basic source of maleness"—perhaps overlooking the biological fact that half of the sperm carry an X chromosome and are therefore "female." According to this article, "supersperm" engage in a "Kamikaze mission," barging ahead to "further the success of their brothers." In contrast, eggs "send alluring chemical clues" and wait quietly to be rescued.[32] Anthropologist Emily Martin observes that such images present sperm and egg as tiny intentional beings—actual selves—thereby "adding to our willingness to focus ever more attention on the internal structures of this tiny cellular self, namely, its genes."[33]

Biological explanations may reassure threatened groups that they possess special skills and advantages, thereby demonstrating their inherent superiority or worth. When feminist texts promote caring or intuition as unique feminine skills, they are effectively depicting their readers as advantaged. When men's movement texts celebrate male aggression as a biological strength, they are elevating a presumed necessary evil to the status of a positive social good. Both groups are engaged in setting boundaries of identity and delineating criteria of social worth. Here as in other forums, genes have become a way to establish the legitimacy of social groupings. This function is even more overt in the public debate over the meaning of race.

Genes and Race

At the 1990 annual meeting of the American Psychological Association, psychologist J. Phillipe Rushton presented a paper arguing that blacks have smaller brains than whites and that this explained differences in their educational performance. Apparently oblivious to the "elephant problem" (if brain size is all, why are not elephants the most intelligent creatures on earth?), Rushton suggested that racial variation in brain size was a consequence of evolutionary pressures.[34] So, too, according to his gene-based theory of racial patterns, genital size and fertility rates vary, as an evolutionary adaptation. As humans entered the colder climates of Europe, the environment favored those who had genes for "increased social organizational skill and sexual and personal restraint, with a trade-off occurring between brain size and reproductive potency."[35]

Rushton's thesis is one variant of the long-standing hypothesis that differences in social, political, and economic status between racial groups are a consequence of biological differences—in brain size, intelligence genes, or other bodily characteristics. The body's supposed differences can be used to explain the social status of oppressed groups in ways that make inequalities appear natural and appropriate. Much of this debate has focused on intelligence and, increasingly in the 1990s, on intelligence genes. But the modern debate on race, genetics, and IQ began in 1969, when Harvard psychologist Arthur Jensen published a famous essay, "How Much Can We Boost IQ and Scholastic Achievement?"[36]

Jensen argued that there was a significant average difference in IQ between blacks and whites of fifteen points on the conventional IQ test. He acknowledged that environment played some role in creating this difference, particularly since the scores of poor, white, urban children were similar to those of poor, black, urban children. But he decided that between one half to one fourth of the mean difference (7.5 points to 3.8 points) in IQ scores between blacks and whites could be attributed to genes. Jensen even used the fact that prosperous blacks tend to have lighter skin than poor blacks as evidence of their genetic superiority (rather than, for example, their greater social acceptability in mainstream white culture). He concluded, as he put it in an interview with *Newsweek*, that the school system needed to learn to accommodate "large numbers of children who have limited aptitudes for traditional academic achievement."[37]

The popular press disseminated Jensen's views. Some journalists supported him as a progressive and authoritative scientist. "Can Negroes learn the way whites do? Findings of a top authority" said a *U.S. News and World Report* headline.

The article warned that Jensen's findings needed to be taken seriously as evidence of genetic differences.[38] But Jensen's claims became, as *Newsweek* put it, a "potential social hydrogen bomb."[39]

Fueling the growing dispute, psychologist Richard Herrnstein published a provocative article in the August 1971 *Atlantic Monthly*. Herrnstein made the ambitious claim that intelligence was 80 percent genetic. He argued that as social programs equalized environmental forces—that is, as all children came to have access to high-quality education, good nutrition, a stable home environment, and cultural opportunities—genetic ability would become even more important in determining levels of achievement. Under such conditions of equal opportunity, inherited ability would be the only variable shaping IQ, success, and social standing.[40] And since blacks were genetically inferior in intelligence, he said, they would suffer the most from a complete meritocracy. *Newsweek* quoted Herrnstein on a point he would develop further in the coming years: "The higher the heritability [of IQ] the closer will human society approach a virtual caste system with families sustaining their position on the social ladder from generation to generation."[41]

Popular coverage in the 1970s was often skeptical, if not critical, of such theories. Herrnstein complained about media bias that made it difficult for his message to be understood by the public. "It takes great pains to get a balanced and reasonably complete account of intelligence testing into the public forum where it might inform public policy."[42] Increasingly defensive, he accused the media of trivializing the research and systematically favoring sociological over genetic explanations.[43] So did Mara Snyderman and Stanley Rothman in their 1988 book on the media coverage of the IQ controversy. They attributed media bias against biological explanations to the "metro-Americans who feel estranged from American values, yet who influence the public through the media."[44] But it was Jensen and Herrnstein—and later Thomas Bouchard (a psychologist at the University of Minnesota)—who, through their publications in popular magazines, appealed to the public as a way to establish the legitimacy of their scientific claims.

In the 1980s, growing concerns about domestic problems—the cost of welfare programs, the changing ethnic composition of major cities, and the growing "underclass"—encouraged speculation about the role of genetics in perpetuating poverty and violence. Code phrases like "welfare mother," "teenage pregnancy," "inner city crime," and "urban underclass" were often indirect references to race. But some public figures did not hesitate to make the connection explicit. Marianne Mele Hall, a Reagan administration appointee, announced in 1992 that African Americans were "conditioned by 10,000 years of selective breeding for personal combat and the anti-work ethic of jungle freedoms."[45] Columnist George Will, in a 1991 *Newsweek* column inspired by a speech by Harvard professor James Q. Wilson, proposed that a black "warrior class" in the inner city was a consequence of nature "blunder[ing] badly in designing males." Men are innately uncivilized, he said, and though socialization has often constrained biology, two "epochal events" have changed this picture: "the great migration of Southern rural blacks to Northern cities and the creation of a welfare state that made survival not dependent on work or charity."[46]

In February 1992 Dr. Frederick Goodwin, Director of the National Institute of Mental Health in the Bush administration, presented a similar view in a statement before the NIMH Health Advisory Council. Goodwin described primate studies indicating that violence is a "natural way for males to knock each other off." This led him to speculate that "the loss of social structure in this society, and particularly within the high impact inner city areas, has removed some of the civilizing evolutionary things that we have built up, and maybe it isn't just the careless use of the word when people call certain areas of certain cities 'jungles,' that we may have gone back to what might be more natural, without all of the social controls that we have imposed upon ourselves."[47] The speech evoked a strong reaction from members of the Black Congressional Caucus, who interpreted his remarks as a racist comparison between inner city youth and monkeys. Goodwin apologized for the comment, but later that year reiterated these views in a speech to the American Psychiatric Association about the administration's "violence prevention initiative," a program that included research efforts intended to identify those people who may be biologically prone to violent behavior.[48]

Genetic language about race appears in a more positive, but no less stereotyped, form in the context of sports. Racial differences in aptitude for sports was a prominent theme throughout the history of the eugenics movement and an integral part of Nazi sports medicine. The theme persists today, though sportscasters have been cautious since CBS commentator Jimmy the Greek was reprimanded for saying on the air that blacks were better athletes than whites because they were bred that way. Their bigger thighs, he said, developed in slave trading days "when the slave owner would breed his big black to his big woman so that he would have a big black kid."[49] References to genetic characteristics appear in a less direct form when television commentators describe African American athletes as having "natural grace," "gifted bodies," or "genetic talent."[50] To black baseball player Reggie Jackson, such flattering images, all referring to traits that are not learned, are also comments about the intellectual abilities of African Americans. The subtle message, he says, is that "we have genetic talent but we're just not intelligent."[51] But the message from Al Campanis, former general manager of the LA Dodgers, was far from subtle. Explaining the lack of minorities in upper management in sports, he said: "Blacks lack the necessities to take part in the organizational aspects to run a sports team."[52]

In the 1990s, race theorists are more and more willing to publicly express their views about genetic differences between ethnic groups and to suggest the significance of such differences for social policy. Physical anthropologist Vincent Sarich at the University of California at Berkeley lectured to hundreds of undergraduates on the genetic basis of racial differences.[53] Michael Levin of City College in New York has not only argued that blacks are less intelligent than whites but also used his theories to oppose affirmative action. He has asserted that differences in average test scores (which unquestionably exist) are self-evident proof of genetic differences—as though the SAT provided direct access to DNA.[54] By 1994, the extravagant publicity launching Herrnstein and Murray's *The Bell Curve*—an immediate best seller—moved the debate over genetic differences to center stage.

For African Americans, such public statements can be both discouraging and damaging. As one minority student put in a *New York Times* interview, African Americans are undermined by assumptions that they are not smart enough to make it, that "we don't have that thought gene."[55] To protest such constructions, some African Americans have proposed a counternarrative drawn from a long history of Pan-African ideology, in which black skin is a sign of superiority. In public lectures, Leonard Jeffries, the professor at City College in New York quoted at the beginning of this chapter, presents a view of world history and race biology that celebrates biological differences. "Black Africans of the Nile Valley," he claims, are the source of all science, mathematics and religion. And melanin, the pigment that makes skin dark, is a crucial biological need. "You have to have melanin to be human. Whites are deficient in it . . . . it appears that the creative instinct is affected."[56] Distinguishing sun people (blacks) from ice people (whites), he is interested in promoting collective identity on the basis of a biological construction of "race."[57]

His ideas are controversial within the African American community, but he is not alone in his goal. Just as some feminists use biological explanations to affirm the special characteristics of women, so many African Americans want to affirm the essential characteristics that define their unique identity, while rejecting stereotypes imposed by others. Some African Americans have supported and sustained the uniquely American idea that anyone with "one drop" of African blood should be classified as "black," for such a social perception is important to civil rights and affirmative action. These African Americans have opposed proposals for a new "multi-racial" category in the next census, arguing that such a category would make it possible for mixed-race "blacks"—who by some estimates make up more than 90 percent of the American black population—to change their racial status. And even if 10 percent of them chose to do so, it could affect legislative districts, school desegregation plans, and regulatory programs concerning housing, employment, and education.[58] The black "race" may be a biological fiction, but it is a politically important reality, perhaps most of all to those who have historically been oppressed by their status as members.

Purity of blood is a frequent theme in literature dealing with racial difference in American society. Dep, the hero of Spike Lee's 1990 film *Mo Better Blues,* chanted during a game of "playing the dozens": "No White Blood in me. My stock is 100% pure." His girlfriend replied: "Master was in your ancestor's slave tent just like everyone else."[59] Novelist Toni Morrison wrote in her 1981 novel *Tar Baby* about the plight of Jadine, a "yalla," suggesting that having large amounts of white blood creates a biological tug-of-war in the psyche.[60] Genetic images appeal to these writers as a way of resisting cultural imperialism and establishing collective identity on the basis of shared identification with a common ethnic heritage.

Ideas about the biological inferiority of whites also appear in this literature. Exploring racial differences, Morrison presents one perspective in her 1977 *Song of Solomon.* Her character Guitar says that "White people are unnatural. As a race they are unnatural. They have a biological predisposition to violence against people of color. The disease they have is in their blood, in the structure of their chromosomes."[61] Her character's comments are reflective of the claims of some leaders of the Afrocentric movement. Michael Bradley's 1991 book *The Iceman Inheritance: Prehistoric Sources of Western Man's Racism, Sexism, and Aggression*—popular with Afrocentrists—argued that white men are brutish because they are descended from Neanderthals.[62]

Afrocentrists are effectively attempting to transform their differences into positive biological strengths. But they share an assumption with racist critics: that race is a biological reality with some meaning for this debate. For gay rights activists, the problems are different; they face the daunting task of redefining a "sin" or a "lifestyle choice" as a biological

118

imperative. In the debates over the "homosexual brain" or the "gay gene," nature and nurture have even more complicated meanings.

The Homosexual Brain

In August 1991, Simon LeVay, a neuroscientist at the Salk Institute, published a paper in *Science* that linked homosexual behavior to brain structure. LeVay said that homosexual males, like all women, had a smaller hypothalamus than heterosexual males. The hypothalamus is a part of the brain between the brain stem and the cerebral hemispheres, believed to play some role in emotions. It is too small to be effectively examined through contemporary brain imaging techniques such as Positron Emission Tomography (PET scans). LeVay needed, therefore, to study the brains of cadavers. His conclusions were based primarily on the postmortem examination of the brains of 41 persons, 19 of them homosexual males who had died of AIDS. He has acknowledged that his findings were open to several interpretations. The size differences in the hypothalamus could indicate a genetic basis of sexual orientation, but they could also be a consequence of behavior, or they could be coincidental, reflecting neither cause nor effect but the presence of some other condition (such as AIDS).

LeVay preferred the genetic explanation, describing his "belief," his "faith" in the biological basis of behavior. Indeed, LeVay's research followed from his personal conviction that "I was born gay." He has stated that virtually all human variation, including detailed personality differences and such cultural preferences as musical taste (Mahler over Bruckner, for example), are biological. LeVay is convinced that children are entirely genetic products. Some children are, from the moment of conception, fated to become gay; if parents have any influence at all, LeVay argues, it is only in the way they respond to the inevitable.

LeVay's claims were later supported by the findings of a team of geneticists led by Dean Hamer at the National

119

Cancer Institute. In 1993 they claimed to locate genes on the X chromosome that predisposed some men toward homosexuality—the X chromosome is inherited in boys, of course, only from the mother.[63] This report and Hamer's popular book on the subject received extensive media coverage and attracted significant public interest.

The response to these claims reflected the long public debate over the social meaning of homosexuality. Homosexuals have at various times been labeled as criminals, sinners, physiologically degenerate or psychologically disturbed; in some societies they have been assigned ritual significance. In a sociology of homosexuality, David Greenberg has traced the notion, recurrent from the Renaissance to the late-eighteenth and early-nineteenth-century phrenology movement, that homosexuality is a biological and innate condition.[64] This remained a minority view, however, overwhelmed by moralizing ideas about sexuality and free will. For most of the Christian era, homosexual desire and behavior have been culturally defined as sins, moral failings, and individual choices that could be changed through prayer, repentance, and will power. Since the late nineteenth century, many physicians have also identified homoerotic attraction as a medical condition to be cured through psychoanalysis, aversive conditioning such as electric shock, and drug therapy. Yet despite its medicalization, homosexuality was (and remains) in many places a crime against the state, defined by anti-sodomy statutes.

In the 1960s and 1970s, gay rights activists began to challenge prevailing ideas about the legal, moral, and psychological meaning of homosexuality. They met with some success, convincing the American Psychiatric Association in 1973 to remove homosexuality from its listing of recognized mental illnesses, the Diagnostic and Statistical Manual.[65] Gay activists were convinced that medicalization promoted homophobia by constructing homosexuals as diseased. Hoping that demedicalization would precipitate social change, they fought to define homosexuality as an innate and therefore normal variation. The changing social beliefs exemplified in demedicalization did have some impact. Some states ceased enforcing anti-sodomy laws. And some religious congregations began to accept homosexuals as active parishoners, even as ministers. At the same time, however, the AIDS epidemic contributed to increased public fear of homosexuality as an emblem of death. The homosexual community was itself transformed by AIDS, and homosexual males faced new forms of discrimination as the "gay plague" invested their activities with new moral and medical meanings.[66]

In this context, the research constructing homosexuality as biological had a tactical advantage; it shifted responsibility from the person to the genes. Individual homosexuals had no choice but to behave as they did. It would therefore be unjust for society to discriminate against them, for the Constitution, demanding equal protection, prohibits discrimination on the basis of immutable characteristics.

LeVay thus sought publicity for his research, and his conclusions became a media event, discussed in popular magazines, major newspapers, and on television talk shows. The hypothalamus, a little-known organ deep within the brain, became a popular symbol of virility. A Calvin Klein advertising campaign referred to a "hypothalamus-numbing host of imitators." A *Newsweek* article called "Born or Bred?" explored the implications of the "new research that suggests that homosexuality may be a matter of genetics not parenting." The magazine's cover photo featured the face of an infant, with the headline "Is This Child Gay?"[67] "Is lesbianism a matter of genetics?" asked the headline of another 1993 article. "Little girls are made of sugar and spice and everything nice, and some of them may have a dollop of genetic frosting that increases the likelihood they'll grow up gay."[68] Vice President Dan Quayle publicly disagreed, however, insisting that homosexuality "is more of a choice than a natural situation. . . . It is a wrong choice."[69]

The debate was joined on television's "Nightline" in a program on homosexuality. The topic was whether "a newborn infant may already have certain physical differences in the brain that could be distinguished from the brain of an infant that will grow up to be a heterosexual." A leader of the religious right, Jerry Falwell, appeared on the show to insist that homosexuality was not innate but a learned and chosen life style; he worried that the research would be used to legitimate homosexual practices. Meanwhile, host Ted Koppel, referring to a "potentially gay fetus," asked: "Will people abort?" (Extending this idea, a 1993 Broadway play called "Twilight of the Golds" featured a geneticist and his wife who learn through prenatal tests that their unborn son will be gay. After much-soul searching, they abort the fetus; the family is torn apart.)[70]

The media also speculated on the potential effect of genetic research on homophobia. On a segment of the prime-time news show "20/20," Barbara Walters asked: "I wonder if it were proven that homosexuality was biological if there would be less prejudice?"[71] *Newsweek* presented the views of homosexuals who welcomed the research, anticipating that it could reduce animosity. "It would reduce being gay to something like being left-handed, which in fact is all that it is," said Randy Shilts.

But the gay community has been divided about the consequences of genetic identification. Some anticipate abortion of "gay fetuses," increased discrimination empowered by genetic information, or the use of biotechnology to control homosexuality, for example with excision of the "gay gene" from embryos before implantation.[72] Janet Halley, a law professor, has predicted that essentialist arguments of biological causation will work against constitutional rights and encourage "the development of anti-gay eugenics."[73] The *National Inquirer* responded to research on the "gay gene" with the headline: "Simple injection will let gay men turn straight, doctors report."[74] And a spokesman for the National Gay and Lesbian Task Force suggested that genetic thinking would give rise to the idea that "by tweaking or zapping our chromosomes and rearranging our cells, presto, we'd no longer be gay."[75]

There is some historical justification for these concerns, since Nazi extermination of homosexuals was grounded in their presumed biological status. And other campaigns by gay activists have had unexpected consequences: The 1973 American Psychiatric Association decision to change the DSM classification failed to produce the social legitimation anticipated by those who had advocated the change. LeVay

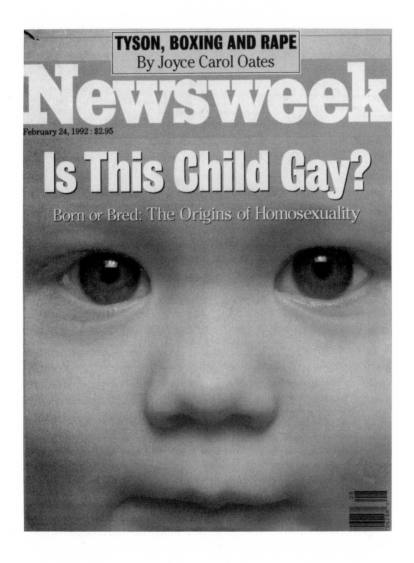

TYSON, BOXING AND RAPE
By Joyce Carol Oates

Newsweek

February 24, 1992 : $2.95

Is This Child Gay?

Born or Bred: The Origins of Homosexuality

himself dismisses such historical precedents: "Those who look to history are condemned to repeat it."[76]

If "Dear Abby" is any indication, however, the biological narrative has influenced popular beliefs about homosexuality. In 1992, when a reader complained about the columnist's suggestion that homosexuality was a consequence of both

123

nature and nurture, Abigail van Buren responded: "I have always believed that one's sexuality is not a matter of choice—that homosexuals, like heterosexuals, are born that way. I apologize for my lapse in judgment in buying that nature-nurture theory. I knew better and am profoundly contrite."[77] This strong statement in such a visible source suggests that in the short term, at least, biological explanations have gained ground in the popular understanding.

Biological Difference and Social Meaning

In the public debates over human differences—for example, the meaning of gender, race, and sexual orientation—genetic images are strategically employed in an effort to delineate boundaries, justify rights, or legitimate inequalities. Genes can be understood in this debate as rhetorical devices that can be utilized in many different ways. They have been used to identify biological differences and to give them social meaning—by those, for example, who believe education will make no difference in the social status of black Americans; by those who favor homosexual marriage; by those who promote equality of the sexes; and by those who oppose equality in general.

Biological differences in themselves have no intrinsic social meaning. Skin color is genetic—it is a real biological property—but it became a sign of political and economic difference for specific historical reasons, including the European colonization and exploitation of Africa. Due to the vagaries of evolution and population genetics, African populations happened to have skin that was uniformly darker than that of European populations. If both Africans and Europeans had instead manifested equivalent variation in skin color (displaying skin tones within each group ranging from very light to very dark), skin color would not have been a reliable sign of Continental origin and therefore could not have served as a visible mark of social or economic status. (Perhaps some other biological trait, such as eye color, would have come to stand for racial difference.) Certainly racial classifications vary across culture. For example, Brazilian ideas about race, the anthropologist Marvin Harris has observed, would be "inconceivable in the cognitive frame of descent rule" that guides American ideas. Full siblings in Brazil can be assigned to different racial categories if they differ in physical appearance.[78]

Sex, too, has a complicated history as a social category. Thomas Laqueur's work has demonstrated that for much of human history, from classical antiquity to the end of the seventeenth century, men and women functioned in two different social roles but were seen as variations on essentially one biological sex. The boundaries between male and female were understood to be "of degree and not of kind." To Galen, for example, the sex organs of both men and women were basically "the same," the uterus seen as a form of the penis, the ovaries a form of the testicles.[79] The biological story of difference was rewritten in the midst of the scientific revolution, Laqueur has argued, and two distinct biological sexes became a political necessity by the late eighteenth century on account of economic and social changes.[80] From another perspective, the biologist Anne Fausto-Sterling has noted that people do, biologically, come in more than two sexual forms—some experts estimate that hermaphrodites (individuals who have some combination of both male and female genitalia) account for one in every twenty-five births, or 4 percent of the human population. These intersex individuals are socially invisible because of medical management: Such infants are promptly designated male or female and their genitals surgically transformed.[81]

The existence of the homosexual body, too, depends on culture. In Greece in the fourth and fifth centuries B.C. there was no culturally recognized distinction between heterosexuality and homosexuality. Greek thinkers found nothing surprising in the coexistence of desire for both male and female sexual partners. They were, however, concerned about the control of desire and the uses of pleasure, and Greek texts devote significant attention to questions of control and power, though virtually none to sexual orientation.[82]

The biological groupings that appear in the contemporary debate, then, are specific historical products, not necessary biological categories. The meaning of these groups as genetically constructed is likewise flexible. Biological differences can become a source of stigmatization or of group pride. They can justify remedial socialization (extra math study for girls) or regressive social policy (expecting all mothers to stay home with their children). They can also be a source of political power (legal recognition of homosexual marriage with all attendant benefits, for example). When defined as an unchangeable and fundamental biological attribute, race, sex, or sexual preference can become a source of social support and authenticity that may be particularly valued by groups that have been the focus of past discrimination.

Biological narratives do not inherently oppress. But we argue that they are dangerous precisely because of the cultural importance attached to DNA. These narratives, attributing social differences to genetic differences, are especially problematic in a society that tends to overstate the powers of the gene. Charged with cultural meaning as the essence of the person, the gene appears to be a powerful, deterministic, and fundamental entity. And genetic explanations—of gender, race, or sexual orientation—construct difference as central to identity, definitive of the self. Such explanations thereby amplify the differences that divide society.

It is especially ironic that DNA has become a cultural resource for the construction of differences, for one of the insights of contemporary genomics research is the profound similarity, at the level of the DNA, among human beings and, indeed, between humans and other species. We differ from the chimpanzee by only one base pair out of a hundred— 1 percent—and from each other by less than 0.1 percent. The cultural lesson of the Human Genome Project could be that we are all very much alike, but instead contemporary molecular genetics has been folded into enduring debates about group inferiority. Scientists have participated in these debates by seeking genes for homosexuality and alcoholism, genes for caring and genes for criminality. This research and the ideological narratives that undergird it have significant social meaning and policy implications.

7

Absolution: Locating Responsibility and Blame

Political debates over the state of the environment, the declining standard of living, the imbalance of trade, urban violence, single mothers, and the "war" on drugs employ a language of crisis and catastrophe. Illegitimate births are a "plague," crime is "out of control," the environment is "inundated" by toxic wastes, the population crisis is a "nightmare," urban centers are "collapsing," and the "epidemic" of divorce is threatening family values. Such rhetoric is a common response to change, and in this climate, locating blame becomes an important cultural activity. Blaming represents a search for certainty in anxious and disruptive situations: if responsibility can be fixed perhaps problems can be controlled.[1] The gene, as an apparently deterministic force, has transformed public debate about the source of social problems.

Social crises can be blamed on supernatural beings, fate, human nature, or the policies of the state. The kinds of explanations that take hold in a society reflect cultural beliefs—beliefs, for example, about the fundamental nature of human beings, the proper relationship of individuals to society, the role of government, and the importance of social boundaries. Such explanations, says anthropologist Mary Douglas, "signify something about the nature of the society which takes them seriously."[2]

Analyzing the social and political considerations underlying claims of responsibility, philosopher Marion Smiley observes that in the United States "causal responsibility and moral blameworthiness have become interchangeable terms of discourse." The American concept of blame, based on moral agency, has been couched in the language of individual free will: Individuals are seen as the source of social problems and personal responsibility as the way to resolve them. This makes sense, she argues, "in a community which has replaced collective responsibility with a more individualized form of accountability."[3]

This emphasis on individual responsibility plays an important role in American political culture. For example, Federal Reserve Chairman Alan Greenspan asserted in 1992 that the recession would end if only individual consumers would start spending more money; conservative critic Charles Murray has argued that poverty would end if only individual poor people would get to work; former Vice President Dan Quayle has suggested that teen pregnancy can be controlled if only individual teens will take responsibility for their actions; African American scholar Shelby Steele has placed responsibility for black poverty not on racism, oppression, or limited opportunity, but on lack of individual initiative: "Blacks must be responsible for actualizing their own lives."[4] And environmentalists have assigned individual consumers significant responsibility for ecological problems because of the choices they make to use, for example, disposable diapers rather than cloth ones.

This emphasis on individual responsibility effectively deflects attention from the broad social and economic forces within which individuals act. If society is simply a collection of autonomous individuals, then responsibility for social progress and problems lies not with political organizations, not with group actions, but with the individual—whether pathological, admirable, deviant, courageous, or dysfunctional. And if individuals are not malleable—and in the political imagination a genetic trait is a trait that cannot be affected by environmental forces—then efforts to change the social environment may be irrelevant. People with problems become, in effect, problem people. This implies that the government can "only do so much" to help those who are "programmed" to be the way they are, since there are severe limits to the efficacy of social intervention.

This construction of responsibility and blame suggests the ideological utility of genetic essentialism in the 1990s, for genetic explanations of individual actions have been incorporated into both popular media and social policy debates in ways that absolve the social order. Genetic deviance, a property of individuals and their DNA, relieves state and society of collective responsibility for the social conditions that foster violence. Genetic explanations therefore appeal to neoconservatives as a way to rail against the liberal, egalitarian theories of the 1960s. They have become a weapon for those opposing social welfare programs for the needy and rehabilitation programs in the jails. Such programs can seem irrelevant if social problems derive from individual biology.

Yet genetic explanations, in this literature, also exonerate the individual, removing moral responsibility by providing a biological "excuse." Genes are agents of destiny: We are victims of a molecule, captives of our heredity. Such explanations have a double edge, for while they shift responsibility to DNA, they create a new biological form of blaming—for the "flawed" parents who pass on "bad genes," for those who knowingly take genetic risks, or for the physicians who fail to offer parents prenatal tests (see Chapter 9).[5]

We begin this chapter by exploring the rhetoric of individual responsibility as it appears in neoconservative

critiques of social policy interventions, for the neoconservatives have used genetic explanations effectively to set the
terms of public discourse. We then link neoconservative values to the biological narrative in debates about how to deal
with criminal violence and how to raise children. In each
case, biological explanations ("it's in my genes")—much like
theological explanations ("the devil made me do it")—locate
problems (and, therefore, solutions) within individuals.

The Limits of Intervention

Seeking to explain the paradox of poverty in an affluent society, many social policy analysts since the early 1980s have
focused on individual pathology. While 1960s liberals blamed
poverty and crime on the economic and social structure of
the society (the word "disadvantaged" conveyed this view),
1980s neoconservatives shifted the burden to pathological or
irresponsible behavior among individuals in the "underclass." Individuals have a responsibility to transcend their
environment, according to these critics, and welfare
programs like Aid to Families with Dependent Children are
flawed because they assume that government intervention
can solve problems for which only the poor themselves are to
blame. This broad critique of the Great Society programs of
the Johnson administration suggested that poverty results
from moral, not biological or social, failures. Like Lady
Catherine de Bourgh, the character in Jane Austen's *Pride
and Prejudice* who visited the poor in order to "scold them
into harmony and plenty," conservative critics have suggested that with sufficient pressure the poor could be forced
to help themselves.

This neoconservative critique of liberalism was laid out in
1984 by Charles Murray of the American Enterprise Institute
in his influential book *Losing Ground*.[6] In one of his chapters, "The System Is to Blame?" Murray attacked the notion
of structural poverty. Social action programs, he wrote, were
a form of "legalized discrimination" that had "created much
of the mess we are in." He blamed government policies for

encouraging socially destructive behavior, asserting that welfare subsidies encouraged poor women to have more babies: "We tried to provide more for the poor and produced more poor instead."[7]

Murray, in his 1984 book, interpreted individual behavior as malleable and affirmed that changing public policy could change the problematic behavior of the poor. He did not look for fundamental explanations of human behavior: "Whether they are responsible in some philosophical or biochemical sense cannot be the issue." But he focused blame on the individual, and his interpretation of social problems (called "victimism" by his critics) became the theme of numerous subsequent neoconservative attacks on liberalism, affirmative action, and the minority groups and "special interests" claiming expanded rights and federal funds.

In their 1991 study of changing attitudes toward poverty, Thomas and Mary Edsall concluded that policy analysts interpreted the problems of crime and welfare and the condition of life for the poor as "the responsibility of those afflicted and not the responsibility of the larger society."[8] And in *The New Politics of Policy* (1992), political scientist Lawrence M. Mead attacked the philosophy of "sociologism"—his term for the idea that behavior can be changed by improving environmental circumstances. Antipoverty policies failed, he said, because they depended on false assumptions that the poor would get ahead if they were given the chance. Rather, he declared, the source of rising crime, welfarism, and declining schools is "dependency and dysfunction"—that is, liberal social programs and the flaws of individuals. And he questioned whether the poor are "victims or victimizers," the exploited or the exploiters of society.[9]

The actual evidence underlying claims about the harmful influence of government intervention is limited and controversial. Andrew Hacker in *Two Nations* offered data to refute many of the common stereotypes about the effects of welfare. He found that about half of all black single mothers are self-supporting and that affirmative action programs have neither significantly increased the average earnings of black men relative to whites nor changed the relative rates of unemployment.[10] Yet public belief in the "exploitative" poor may not depend on data but on the utility of this construction of the poor in a climate of economic uncertainty.

Mead's criticism of entitlement programs for their failure to deal with the "serious behavioral problems of the poor" was cited in more than 200 syndicated columns. His account became authoritative, a source used to buttress attacks on proposals for tax credits or job programs and to place greater responsibility on the shoulders of welfare recipients. Columnists suggested that poverty is a moral failure and treated the poor as morally blameworthy. Playing on the compelling value placed on self-reliance in America, their message was that social interventions like welfare and affirmative action help to create pathologies in the underclass and to exacerbate social problems.

The neoconservative critics assume that social policies encourage pathological behavior, and their goal has been to eliminate welfare programs. "Victimism," focusing on the individual as the source of social problems, has established a climate of public discourse in which genetic essentialism can flourish. Indeed, the values of the conservative right are shaping biological narratives in contemporary culture. This has become increasingly explicit over the past decade in both popular and professional debates about the causes of crime.

Locating the Blame for Crime

In 1981 President Reagan, speaking to a group of police chiefs, blamed crime on the utopian belief that social programs could prevent criminal behavior: Hindering the swift administration of justice was "a belief that there was nothing permanent or absolute about any man's nature." America was unable to control crime, said Reagan, because it was in the grip of the idea that man was a product of his material environment and that by changing that environment we could permanently change man and usher in a great new era.[11]

Ten years later Daniel Koshland, editor of *Science*, promulgated the same view in an editorial on the pitfalls of looking to childhood experience as the explanation of criminal behavior. He was explicit about the causes of deviance. It was time, he wrote, to recognize that "the brain can go wrong because of hereditary defects that are not related to environmental influences. . . . People believe that the brain can only be affected by the environment; they refuse to face the fact that the brain, like other organs, while functioning correctly in most areas, may have one part go awry and cause major malfunction."[12]

Two years later, in his 1993 book on the evolutionary basis of moral behavior, James Q. Wilson assumed the existence of an intuitive moral sense that is biologically driven; the fact that most people are not criminals, he said, has less to do with social circumstances than with their innate moral sense. He too dismissed the tendency to blame social problems on environmental circumstances, crime on social barriers, educational failures on schools, or promiscuity on television role models. We have only increased social problems, he said, by "replacing belief in personal responsibility with the notion of social causation" and by trying to solve problems such as crime through social programs.[13]

The interest in such reductionist explanations builds on worries about crime and disorder, and on the widespread feeling that social programs have so far failed to resolve the troubling problem of violent crime. The Bush and Reagan Presidential campaigns politicized the growing fear of crime in the effort to discredit Democratic challengers. During the 1980s, policy debates about the causes of crime began to focus less attention on the social circumstances of criminal behavior and more on its psychological and biological basis, shifting blame from society to the individual.

Research on criminal behavior has generally been eclectic in approach, and the discipline of criminology has long included both biological and sociological research. According to David Garland, the modern discipline of criminology indeed developed in the nineteenth century from the convergence of two quite different traditions: scientific interest in understanding the root causes of crime by identifying criminal types, plus the policy-oriented interest in management and control that drew on sociological perspectives.[14]

In the 1950s and 1960s sociological theories prevailed. Criminologists, often writing in the popular media, strongly rejected theories based on biological determinism. A physician who was director of the Bureau of Mental Health Services of the New York City Court of Domestic Relations wrote in a 1954 *New York Times Magazine* story that society—particularly the Cold War—was to blame for the rise in delinquent behavior: "The positive correlation between the rate of delinquency and war and cold war cannot be ignored."[15] Similarly, in a *New York Times* story in 1961, Frederick Wertham, a "psychiatrist in the criminal field," explored the question of why people commit murder. "Violence is as contagious as the measles," he said. He anticipated that social change could make murder obsolete, just as (he believed) it had made incest obsolete.[16]

A decade later the receptive public response to research on the so-called "criminal chromosome" (see Chapter 5) suggested a shift in popular beliefs. The growing tendency to blame criminal behavior on biological predisposition reflected disaffection with the social policies founded on environmental determinism. Attention has increasingly focused on the biological attributes of criminals, sometimes in terms that recall the nineteenth-century Lombrosian focus on "criminal types."[17] Advances in the neurosciences and in brain imaging techniques have encouraged renewed interest in the nature of the criminal brain, and contemporary scientists who study the brain search for individual variations to explain pathological behavior.[18]

Once a taboo topic, the genetic basis of violent behavior has, in the 1990s, become a crucial dimension of both the professional and popular discourse on crime. After the 1993 sentencing trial of James Wanger, a boy who had strangled a friend in a ritual murder, his lawyer explained to the media that a "mutant gene" might have affected his client."[19] Around the same time, a *Washington Post* comic strip by Berkeley Breathed portrayed a young black man beating up another while announcing: "Hey, it wasn't me. It was my *genes.* Experts now say that most anti-social behavior—smoking, alcoholism, drug abuse, violence, even criminality—can all be traced to faulty genes. It's important news. We're not responsible anymore! We're *all* victims. I hurt, too, brother."[20]

In the context of "victimism," genetics is a useful construct, a convenient and apparently scientific way to abdicate responsibility by blaming the person predisposed to violence, the individual genetically flawed. Tied to ideological agendas, biological explanations of violence appeal as a way to protect existing social categories and policies while promising control of those who are defined as a threat to the current order.

As we documented in Chapter 5, many popular accounts in both news reports and fictional narratives presume that certain people, "born bad," will naturally turn to crime regardless of positive social circumstances. Thus, for example, the advantaged twins in the TV movie "Tainted Blood" were predestined to kill because of their genes. Such stories imply that the management of social problems must be based on a recognition of the immutability of human nature, that is, the "fact" that some people are fundamentally evil, driven by their genes. A 1984 book, *Born to Crime*, laid out the dynamics of this policy perspective. Its author, Lawrence Taylor, argued that the law must come to terms with genetics and the ability to predict and treat violent behaviors based on genetic characteristics: "If we choose to accept cancer as being divinely caused, we are unlikely to find a cure for that disease. If we choose to believe that crime is caused by environmental conditions because that theory is socially acceptable, we are unlikely to discover ways to prevent criminal conduct."[21] Echoing this view in a 1992 speech before the American Psychiatric Association, Frederic Goodwin, then director of the National Institute of Mental Health, advised that policies to combat crime should focus on "the individual risk factors that may be important in identifying and treating those who are likely to engage in violent behavior."[22]

The appeal of biological explanations of crime, addiction, and antisocial behavior grows with the apparent intransigence of many contemporary social problems and with public disillusionment about the efficacy of political and social reforms. In fact, whatever the influence of genetics, biological predispositions are not necessary to explain why an inner-city African American, growing up in a climate of racism, drug abuse, and violence, and without much hope of escaping that climate, might become indifferent to human life. Between 1980 and 1990 the federal government withdrew $260 billion from direct support of inner-city communities. Loss of jobs and a declining tax base weakened the community infrastructure.[23] Yet it may be expedient—to those concerned about the public costs of such programs—to believe that problems rest less with society than with inherent predispositions; claims about "predisposition" absolve the political system and blame the individual.

A claim about predisposition after the fact—for example, that a man committed a crime because he was predisposed to criminal behavior—has the rhetorical advantage of being both unfalsifiable and irrefutable. No experiment could disprove the existence of such a predisposition, for the outcome is "proof" of its own cause. By employing this sort of circular argument, genetic predisposition can be invoked to explain any behavior; an individual will respond either poorly or well to a particular environment depending on preexisting biological tendencies. Some have "built in resistance"; others have "evil in the genes."

The concept of genetic predisposition appears to explain a wide range of personality traits and behavior patterns. In childcare books and parents' magazines, for example, authors are now stressing that parents can only do so much for children, since genes significantly determine their behavior and future achievement. In this popular genre of parental advice, predisposition helps to construct a philosophy of limits.

Responsibility for the Child

In 1954, an article in *Parents* magazine described modern parents who were destroying their children by overindulging them. "It has become the fashion to blame everything from arson to kleptomania on an unhappy childhood. But too often parents seem to feel that overindulgence, lack of control, and the granting of a child's every wish will make him happy," said the writer. "Who comes first? Too much self-sacrifice can wreck a home and the children growing up in it."[24] Parents were responsible for establishing a proper balance, for what they did would shape the future of their children.

Thirty-two years later, in 1986, the same magazine described a woman who had grown up as a neglected "latchkey" child. Nevertheless, she earned good grades in school, never stole or used drugs, and never got into trouble. When she had her own child, she did everything she could to provide him with an optimal environment. Yet he became "a drug addict, a thief with no morals and no conscience." The story concluded that "the environment provided by parents actually has very little influence on broad personality characteristics."[25] The article's author supported this view by citing research in genetics that, he said, showed that parents are not necessarily to blame for their children's flaws, since about 40 percent of personality differences between children are caused by heredity over which parents, once they give birth, have essentially no control. This, the author noted, relieves parents of guilt, for even a good parent can have a bad kid. The message of this and many other advice columns written in recent years was summed up in a 1987 *Time* article: "Parents should be blamed less for kids who have problems, and take less credit for kids who turn out well."[26]

Although advice books and parenting magazines are very widely read, they may have limited influence on how parents actually treat their children.[27] Their changing standards, however, do reflect popular sentiments and ideas about child rearing. The parental advice literature in the 1950s was, of course, dominated by the words and works of Dr. Benjamin Spock. The single most important purveyor of child rearing advice in postwar America, Spock has influenced generations of parents. His original (1945) edition of *Baby and Child Care* went through 179 printings before the 1968 revised edition, making it the best-selling new title issued in the United States since 1895, when best-selling lists were first compiled.[28] Spock popularized Freudian understanding of childhood fears and sexuality and has, with relative consistency, followed a psychoanalytic model of human personality development. Like most of the parent advice literature of the postwar period, Spock emphasized the crucial role of environment in shaping the child. A mother's attitudes and decisions were depicted by Spock as having drastic effects on her children's personalities and development. Seemingly minor parental errors, according to this model, could have terrible consequences.

Consistent with the social policy perspectives of the 1960s and 1970s, other child rearing experts of those decades also emphasized the crucial role of the social and family environment in shaping the child. Children with behavior problems, for example, were diagnosed as having "working mother's syndrome." Child psychiatrist Irving Harris attributed all learning problems to home life: "In the home, most of the reasons for failure in learning and success in learning will concern the mother," who might cause problems if she were ambitious, jealous, tyrannical, divorced, or emotionally unstable, but most of all if she were to "enjoy working outside the home."[29] So pervasive was the environmental model during this period that a 1961 article in *Family Circle* asked plaintively: "Are Parents Always to Blame?" The author observed that "many parents, feeling completely responsible for their children's character and personality, reproach themselves with 'How have we failed?' or 'What did we do wrong?'"[30]

By the 1980s, however, childcare books and magazines were appealing to biological determinism as a way to limit parental responsibility. For example, the cover blurb of Stella Chess and Alexander Thomas's 1987 book *Know Your Child* read: "At the heart of the book is the authors' important finding that children have recognizable individual differences or temperaments right from the start. They show how parents can avoid problems by matching their own expectations and attitudes to their child's temperament."[31] In 1993, *Child* magazine described *Know Your Child* as a "breakthrough book" because it had helped so many parents realize that they must work within their child's "innate personality."[32]

Even Spock included a caveat in his 1985 edition: "There is considerable evidence that different children are born with quite different temperaments." He advised parents to accept their children as they are: "One gentle couple might be ideally suited to raise a boy with a gentle, sensitive nature . . . [but] not nearly so ready for an energetic assertive boy. Another couple may handle a spunky son they call Butch with ease and joy but be quite disappointed with a quiet one." Parents can't order what they want, Spock advised. "They take what they get."[33] Despite Spock's continued general commitment to a psychoanalytic approach, the latest editions of his book, revised with the help of colleagues, have reflected the increasing popular and professional interest in biological explanations for children's behavior.

The parental advice literature in this genetic age conveys a message of fatalism. Responsible parents must accept the limits imposed by nature; they must work within their child's "biological potential." A 1991 article in *Parenting* put this bluntly: "The very idea of mentally stimulating a baby is misguided at best and harmful at worst. . . . Intelligence can't really be improved. . . . Your baby has a developmental schedule all her own, and no class can speed her up." If they've got the genes they've got it. If not, don't fight it. Overstimulation beyond genetic capacity, warned the author, can affect a baby's sense of self-worth.[34] A similar fatalistic message shaped the advice offered by Peter Neubauer, M.D., in *Nature's Thumbprint* (1993): "Children are born with a certain capacity for intelligence, talents, to some degree character. . . . They cannot be produced through wishes and expectations."[35] A 1991 article in *Child* advised: "You can enhance a baby's potential but you can't turn her into a genius if she hasn't got the genes for it." The author went on to tell her readers that "Nature has moved to center stage," and she promised that research will give us an "early peek at genetic endowments."[36] And *Parents* magazine, speculating on "Why Kids Are the Way They Are," emphasized the importance of recognizing limits. Environment can modify the "script" submitted by genetics, but parents must "work within the framework of [a] child's nature."[37]

Explaining achievement, intellectual potential, and social skills in reductionist terms can lead to fatalistic preoccupation with identifying and defining limits. Parents are told that they must monitor their children in order to determine their potential and their limitations—to decipher the "script" imposed by their genes. In a 1988 article called "How healthy is your family tree?" a medical journalist advised that everyone should make a genogram with all the information available about relatives as far back as they can be traced.[38] *Child,* in an article called "Cracking the Code," advised mothers to track the brain development of their babies to discover their inherent capabilities.[39] Stanley Greenspan's 1985 baby care book *First Feelings* warns parents that genetics "lays the track" for the train of development; like many such advice books, it lists what the normal baby should be doing at every age.[40]

This parental advice literature draws heavily on the work of psychologists Jerome Kagan and Sandra Scarr. Kagan, observing a correspondence between shyness and inherited variation in the physiological characteristics of the brain, speculated that "children we call inhibited belong to a qualitatively distinct category of infants who were born with a lower threshold for limbic-hypothalamic arousal to unexpected changes in the environment."[41] Watching how two-year-old girls reacted in unfamiliar surroundings, he found that some seemed immediately at ease, while others withdrew. Kagan described his research not only in professional journals, but also on "The Donahue Show": "A small proportion of children, we think 10 to 15%, are born with a slight push from nature to be either outgoing . . . or with a slight push to be fearful."[42] And in 1994, he published a popular book, *Galen's Prophecy,* speculating that innate temperaments affect marital choices, criminality, ethical stances, psychopathology, and parenting.

Kagan has been defensive about his findings, suggesting that as a society Americans are reluctant to acknowledge that biological differences exist among children because it violates egalitarian ideologies. Yet the response to his work was hardly "reluctant." It has been extensively and enthusiastically covered and broadly generalized in the mass media. A

1988 *American Health* article reported that children are "Born Bashful."[43] And *Parents* magazine announced that shyness is part of the "basic temperament."[44] In Kagan's many popular articles, he suggests the reason for the appeal of his findings: genetic explanations deflect blame from mothers. He stated this bluntly in an interview in a *U.S. News and World Report* article called "Your Mother Did It To You," telling the reporter that mothers are wrongly held accountable for their children's successes or failures. "We have attributed all the influence to her as if she were a sorceress. She's a good witch if the child turns out fine and a bad witch if [he or] she doesn't."[45]

Sandra Scarr, a former student of Arthur Jensen, has taken the idea that genes determine a child's personality and eventual success in life even further than Kagan. Scarr has advanced the proposition that children are born with personality characteristics that are so predetermined that "being reared in one family rather than another, makes few if any differences in children's personality and intellectual development."[46] Recognizing this, she said, "might help alleviate needless sacrifices and emotional turmoil." Because children are preadapted to respond to a specific range of environmental opportunities, variations within this normal range are functionally equivalent, she argues. Differences among home environments have, therefore, little effect on differences among children. Scarr advanced this as a "comforting idea [that] gives parents . . . a lot more freedom from guilt when they deviate from culturally prescribed norms. . . . Outcomes do not depend on whether parents take children to the ball game or to a museum so much as they depend on genetic transmission." Scarr too has become a media scientist, helping to organize and appearing in the PBS television series "Childhood."[47]

Genetic explanations appear to be especially liberating to mothers at a time when the meaning of motherhood is a source of social tension and public debate. In 1990, 53 percent of those mothers with a child less than a year old were in the labor force, as compared to 31 percent in 1976. And as more and more mothers of young children work outside the home, they leave the day-to-day care of infants and preschoolers to others. Parents today spend 40 percent less time with their children than they did 25 years ago, and up to one third of school-age children are "latchkey kids."[48] When most mothers stayed home they were blamed for virtually every problem of their children. Now that most mothers do not stay home they are also blamed, as the popular press debates guilt-inducing questions about "What happens to infants when their mothers go to work?" and scientists report on the problems of "maternal deprivation" or the perils of failing to "bond."[49]

For guilt-ridden parents, genetics provides a convenient response. If a child's behavioral problems are inborn, this seems to absolve parents from blame for how the child "turns out." A *Wall Street Journal* story reporting that Type A mothers tend to have babies with the same personality trait evoked a frank remark from the writer, herself a working mother: "This takes some of the weight off me. . . . I'm up for anything that says it's not environmental."[50]

Some current advice books explicitly target the "blame the mother ideology," using the concept of predisposition to reduce maternal guilt. For example, Chess and Thomas began their *Know Your Child* by describing women who blamed themselves for their children's troubling behavior. "Why am I such a bad Mother?" asked one women of her pediatrician, who responded that the child may have been born with a bad temperament.[51] Similarly Elaine Fantle-Shimberg, in *Depression: What Families Should Know*, told her readers to "stop blaming yourself for anything. . . . It's unlikely the child's depression comes solely from something you did wrong." Rather, it is more likely that the youngster "had a vulnerability to depression."[52] *Newsweek* conveyed the same message in a story about a depressed woman who had long blamed her mother for her unhappiness. Her mother had never done the kinds of things (baked cookies) that she was supposed to do. But with new and better knowledge, the daughter placed blame for her depression on her biological roots.[53]

The use of biology to deal with parental blame and guilt has also played a role in Hollywood. In the 1989 film *Parenthood* a couple, meeting with a school psychologist to discuss

142

the troubling behavior of son Kevin, blamed themselves for failing him. But the psychologist reassured them that the problem was "from the womb." It was "chemical." The father, relieved to hear that Kevin's problems were not the fault of his parenting practices, was, significantly, also troubled: if he had another child, would he pass on these bad qualities again?[54] His distress captured the double message in the biological thrust of recent parenting advice. Genetic determinism can relieve the sense of parental responsibility and blame, but it may also create another, even deeper, level of parental guilt—the guilt of passing on bad genes.

In the film *Lorenzo's Oil*, based on a true story about a five-year-old boy diagnosed with the rare disorder adrenoleukodystrophy (ALD), this assumption of genetic guilt played an important role. ALD is a horrible degenerative disease that the boy had inherited through his mother's family. In the movie, a doctor told Lorenzo's parents they had lost a "genetic lottery" and that their son would soon die. But the parents relentlessly sought a cure. Micaela, the boy's mother, was especially desperate. As the carrier of the genetic disease, she was driven by guilt to an extreme, self-punishing level of dedication, as if to expiate the sin of her "poisoned blood." As a discerning critic put it, "the ferocity with which she fights for his dwindling life is fueled both by her desire to reverse the biological havoc her genes have wrought and by the impulse to punish herself, to atone for her guilt with suffering."[55] As such a story suggests, parental guilt can take many forms.

The prescriptive literature on parenting serves as a social barometer, revealing beliefs and expectations that shape how parents in contemporary culture are expected to think about their responsibilities to their children. Genetic essentialism in this literature constructs the child as a predetermined packet of genes, not a being to be molded into a productive citizen, but one to be passively and fatalistically accepted. The role of parents is to recognize genetically determined temperaments and to act accordingly to help their children cope. There are intrinsic limits to parental intervention, this literature suggests. At the same time, the genetically constructed child can relieve parents of personal responsibility and guilt. Though genetic essentialism promotes the bodily guilt of passing on "bad genes," it exonerates the parent of the moral guilt of wronging the child by intentional acts. Genetic explanations have become, in effect, a way to cope with blame and responsibility.

Exoneration

In 1974, a philosopher predicted that "one day science will tell us about all the causes of human behavior and when that day comes, we will no longer be able to hold people morally responsible for their actions."[56] The growing use of biological defenses for personal behavior suggests the appeal of this idea. In 1990 Richard Berendzen, the former president of American University, was fired when investigators discovered that he had been making obscene phone calls from his office. His lawyer defended him by citing research revealing that genetic factors can predispose individuals to behave in abnormal ways when under stress. Berendzen, he said in a media interview, should not be blamed for his behavior; genetically predisposed to obsessive compulsive behavior, he was not responsible for his actions.[57] But a subsequent letter to the editor of *The Washington Post* suggested the implications of such a defense. "We now have a new model of exoneration. It is shiny, scientific, and designed for the guilt-free 80s and 90s: nature made me do it. The beauty of this excuse . . . is that it is highly adaptable to middle-class malefactors whose white-collar crimes cannot possibly be blamed on a wretched environment."[58] (Interestingly, the 1994 edition of the Diagnostic and Statistical Manual of Mental Disorders, published by the American Psychiatric Association, includes "Telephone Scatalogia" as a diagnostic criterion for identifying sexual disorders.[59] Berendzen himself, in a later autobiography, attributed his actions to childhood experience of sexual abuse.[60])

The oppressive consequences of focusing blame for social problems entirely on individuals—their diet, exercise, and other personal habits—have enhanced the appeal of genetic explanations.[61] The notion of biological predisposition can relieve personal guilt by implying compulsion, an inborn inability to resist specific behaviors. Biological explanations deflect attention away from the social and economic circumstances that may drive people to violence, depression, overeating, or drink, but they also provide an excuse for those who, driven by their predispositions, their irresistible biological drives, need not blame themselves.

Genetic defenses appear in many forms. When LeVay and Hamer defined sexual orientation as an innate condition and not a matter of choice, they were convinced that genetic explanations would reduce blame and encourage greater tolerance. A 1992 article in the *New York Times* describing obesity as a genetic condition suggested that fat people should not be blamed or made to feel responsible for their inability to lose weight.[62] A writer for *Allure* magazine debunked the myth that exercise can change the body's shape. "It comes down to whether you picked your parents well. . . . The myth that you can change dramatically puts women in the position of thinking it's their fault."[63] The cover blurb for Lionel Tiger's 1992 book *The Pursuit of Pleasure* promoted it as "a real guilt buster." Pleasure, according to Tiger, is an evolutionary entitlement. We enjoy sex, food, travel because biologically we need pleasure for survival. Individuals, therefore, do not have to be blamed for their foibles, because their foibles are in the genes.[64]

By 1994, Tiger's message was becoming fashionable. "Eat, Drink and Be Merry is Now In" was the lead in a *New York Times* article describing the trends of the new year.[65] The reporter observed that people are giving up on the quest for a fit, healthy, aerobicized life in favor of the sybaritic. The market is expanding for "don't help-yourself" books such as *What You Can Change and What You Can't Change* written by Martin Seligman, Professor of Psychology at the University of Pennsylvania.

Most consumers of advice books and policy guides are middle class, born into relative privilege. For them, the Horatio Alger myth, the "bootstrap" ideology, the idea of overcoming obstacles in their rise to economic success, can hold little of its traditional meaning. But the contemporary advice literature offers a vision that reconciles the Horatio Alger myth with the new story of genetic determinism. We may not be victims of poverty, but we are victims nonetheless—of traumas, addictions, or genetic predispositions. In many of these narratives, however, biology does not necessarily mean destiny. Something, the stories tell us, can be done.

Many child care articles, for example, begin by stressing genetic predispositions and conclude by laying out ways to adapt. This advice, often appearing in the form of a morality tale, defines new responsibilities that include predicting children's "predispositions" and providing them with appropriate therapy. In a *Psychology Today* article called "Arresting Delinquency," the writer told the story of a young man's potential for criminality and how that predisposition was detected and averted by aggressive intervention. "By the time he was 12 there were clear signs that Clifford could become a delinquent, possibly a felon or even a career criminal." But his mother sent him to a clinic where he was tested, then given medication and a behavior modification plan. Genetic testing of children is a way to target those at risk, noted the author. It is "responsible" parental behavior.[66]

According to one biotechnology firm's advertisement, it is a family's responsibility to store genetic data. Vivigen Genetic Repository, a gene data bank that stores family cells, suggests that if you fail to donate your cells, your family may be excluded from the benefits of future genetic tests: "Here's a health gift that really lasts: DNA for your loved ones for future genetic analysis."[67] In a story called "The Age of Genes," *U.S. News and World Report* reported that "advances bring closer the day when parents can endow children not only with health but also with genes for height, good balance, or lofty intelligence."[68] An article in *Mademoiselle* magazine told its readers that they can fix their genetic flaws: "Just because this giant slot machine spits out the blueprint for all your bodily traits doesn't mean that you have to grin and bear those features that you don't fancy."[69] And an advertisement aimed at the athlete who wishes to increase muscle growth inquired: "Bad Genetics?" You can buy "Advanced Cell Growth Formula" to "optimize genetic potential."[70]

These messages suggest that individuals are responsible for knowing their predispositions and for making appropriate compensations. They may have to move to a state with a climate less conducive to cancer, pursue a career more appropriate to their limits, or choose a diet that takes into account their genetic vulnerabilities. "You can change your inside programming," said Leo Buscaglia, the best-selling writer of self-help books, if you take "responsibility for determining your own life."[71]

Stories of the genetic causes of behavior can, however, construct individuals as culpable in other ways. Representatives of disability groups have been ambivalent about genetic explanations. They hope that the discovery of genes for depression and other mental conditions will reduce the social stigma of being different. But they are also concerned that such discoveries could affect the reproductive rights of the mentally ill. The National Alliance for the Mentally Ill (NAMI), a grass-roots support group of about 140,000 parents and patients, has taken a significant interest in the research on the genetic basis of mental illness. Mental illness has long been blamed on stress and family pathology, with parents viewed as largely responsible. Many involved with NAMI welcomed genetic explanations, believing they would help destigmatize mental illness and take an enormous burden off families. As more patients have become active in the organization, however, and as it has strengthened its ties with other disability groups, NAMI's official position has shifted. There is growing awareness among NAMI's members that genetic explanations could devalue the mentally ill by labelling them as intrinsically flawed; they fear that belief in genetic causation will limit their reproductive freedom. While families would be relieved of social blame for pathological behavior, they could be blamed in a different way, for passing on "bad" genes.[72]

Beliefs about blame and responsibility have social policy consequences, for placing blame is an implicit call for action—or inaction. Definitions of responsibility can encourage coercive policies, determine the allocation of funds, and shape the nature of social reforms and institutional policies. When responsibility for complex behavior is located in the genes of individuals, this can ratify governmental inaction (intervention will do no good). But it may also justify the discriminatory use of predictive genetic information, as institutions seek to avoid seemingly inevitable and costly problems by anticipating potential risk. And it could justify the political control of reproduction as an alternative to social control. Some of the problematic implications of such beliefs are already apparent in institutional practices that exploit the concept of biological predisposition.

8

Genetic Essentialism Applied

In his 1992 suspense novel *A Philosophical Investigation,* Philip Kerr describes a twenty-first century society ravaged by an epidemic of violent crime. Social planners, who in 2013 had given up on environmental controls after the "fail-ure of schemes which claimed to ameliorate the environ-ment," now focus their efforts on identifying those individuals biologically predisposed to violent acts. All males carry identification cards with their DNA profiles and are expected to participate in the "Lombroso Program" that will screen them for innate predispositions. Lombroso ("Local-ization of Medullar Brain Resonations Obliging Social Orthopraxy") Clinics employ a scanning machine resembling that used in Positron Emission Tomography to identify men whose brain lacks a Ventro Medial Nucleus (VMN) and who are, therefore (according to the conceit of the novel), likely to become criminals. The idea is to track the individual offender "before he offended at all."

In Kerr's society, socially responsible citizens willingly submit to the "public health" screening, which although not "mandatory" is a condition for employment and insurance. Some criminals manage to circumvent this system, however. To avoid detection through DNA fingerprinting, for example, rapists wear condoms. And despite scrupulous government efforts to maintain the secrecy of the Lombroso records, one man identified as VMN negative (his code name is Wittgenstein) breaks into the computer network and expunges his own name. He simultaneously obtains the names of other VMN negative "suspects" and, acting out of logical necessity, proceeds to assassinate them. Murder is "the logical extension of the Lombroso program." And so too is his fate: When captured, he receives the ultimate punishment of the time, PC ("punitive coma"), in the name of public health.[1]

Kerr's fictional scenario draws on the ideology of genetic essentialism and, in Orwellian fashion, caricatures its implications. Although set in the future, his imagined world of testing and control is grounded firmly in some practices of the 1990s, when assumptions about genetic determinism and the predictability of inherent traits are already playing a consequential role in court rulings, school policies, insurance decisions, and employment practices. Institutions establish the categories and classifications through which the individual functions in society, and their policies are increasingly grounded in assumptions about the deterministic power of genes. Courts are drawing on genetic concepts in custody disputes; judges are citing biological expertise in sentencing decisions; insurers and employers use genetic information in efforts to control medical costs; and schools look to individual disorders presumed to be genetically based to explain learning problems.

As a broadly accepted science-based concept, genetic essentialism has become a resource for many institutions, helping them resolve ambiguous and difficult problems. Institutions routinely gain legitimacy by grounding their policies and practices in what are taken to be natural categories.[2] Scientifically sanctioned, such categories are assumed to be impersonal, rational, and value-free.[3] At the same time, popular images of the gene make the use of genetic information socially acceptable. Media images channel public perceptions and help define "normal" social relationships. They frame ideas about appropriate behavior, thereby facilitating these institutional uses of genetic information.[4] The popular powers of the gene seem to promise that DNA, if comprehensively known and accurately understood, could explain both past performance and future potential. Individuals, moreover, are less likely to challenge practices that conform to mass expectations and accepted cultural beliefs.

Popular perceptions of genetics—the assumptions of genetic essentialism—are played out with practical consequences in many contemporary institutional settings. They serve as a basis for important decisions about family relationships, influence legal interpretations of criminal culpability, and help institutions to anticipate future risk by identifying individuals who are predisposed to health or behavior problems.[5]

Solving Family Disputes

In popular scenarios of the "molecular family," the genetically bound family appears to be more "solid" than the family held together by shared experiences or common values (see Chapter 4). And, in legal disputes over the custody of children, genetic ties increasingly take precedence over emotional or social ties. Judges are invoking genes to explain their decisions and interpreting genetic connections as central to personal identity. This represents a shift from the long-standing goal in family law of assuring children continued contact with the significant people present in their lives.

For example, adoption laws originally were shaped by social concerns for the integrity of ongoing family relationships. To create a shield of privacy between the family and society, the law fostered a legal fiction of natural birth, and in most jurisdictions original birth certificates and adoption records were permanently sealed.[6] Similarly, most courts have adjudicated custody disputes on the basis of judgments about the psychological best interests of the child. The goal has been to preserve "existing affection relationships."[7] Guided by psychoanalytic theory, custody procedures have considered emotional ties a critical factor in decisions involving children. Courts, for example, have intervened in the biological family in order to protect the child's emotional health; they have avoided disrupting emotional bonds between children and their "psychological parents."

When the person is reconceptualized as a genetic entity and forging genetic relationships becomes a goal in itself, however, legal protection for social interests and emotional ties is weakened. Encouraged by the popular belief that genetic relationships are the essence of individual identity and that "knowing one's roots" is a prerequisite to becoming a functional adult, many adoption agencies have now opened birth records.[8] Twenty years ago, adoption agencies discouraged adoptees from searching for their birth parents. Today, confronted by increasing requests to locate genetic relatives, they facilitate searches for "natural" parents and "biological roots"—though such searches may dismay the adoptive family and violate the biological family's interest in confidentiality. These new practices rest on the belief that knowing genetic connections is so critical to a person's identity and emotional health that it overrides norms of privacy and confidentiality.

Recent custody disputes also demonstrate the increasing legal sympathy for genetic ties and the use of such ties to explain legal decisions that could be justified in other ways. The decision in the 1990 California dispute *Johnson v. Calvert* over a child born to a surrogate mother provides a graphic example. The Calverts had donated their gametes (both egg and sperm) and contracted with a surrogate, Anna Johnson, to carry the fertilized egg to term. Press accounts called it a case of "genetics vs. environment"[9] when Anna Johnson refused to relinquish the baby, Christopher, at his birth. DNA tests confirmed that the Calverts were Christopher's genetic parents, and the judge awarded them sole custody.

The significance of this decision for our purposes lies in how it was framed. The judge could have simply enforced the terms of the original contract. He could also have used the traditional standard of the child's best interest and weighed Johnson's claim as a single mother against the value of placing Christopher with two parents who could better provide for him. Alternatively, the judge could have considered the biological processes of pregnancy and birth as a legitimate basis for a motherhood claim and awarded custody or visitation rights to Anna Johnson. Instead, the judge linked the child's proper place to his genetic lineage and justified his decision in terms of the child's genetic endowment:

We know more and more about traits now, how you walk, talk and everything else, all sorts of things that develop out of your genes, how long you're going to live, all things being equal, when your immune system is going to break down, what diseases you may be susceptible to. They have upped the intelligence ratio of genetics to 70 percent now.[10]

The judge noted the "tremendous need out there for genetic children,"[11] and cited popular beliefs about the importance of genetics. He called Anna Johnson, the surrogate mother, a "genetic hereditary stranger" to the child. In effect, he defined the child as a packet of genes and assumed that shared genes are the essence of identity and the basis of human relationships.[12]

Assumptions about the importance of genetic relationships are also appearing in legal disputes between genetic and adoptive parents. In a 1989 case, *Coburn v. Coburn*, a divorced father petitioned for visitation rights with the daughter he had nurtured for ten years. The court dismissed the petition when genetic tests revealed that the child was not his biological daughter. Although the decision was later reversed on appeal (the appelate court considered the effect of "chaos on the emotional well being of the child")[13], several justices still adopted a biological perspective, declaring, for example, that "knowledge of one's biological parents and hereditary history is crucial in ordering one's affairs and making life's decisions."[14] The court, in approving DNA testing, took the position that the child had a right to know her genetic background, but it did not consider the danger that this genetic information might pose to her relationship with the man she had long considered to be her father in her day-to-day life.

Regardless of how a case is resolved, courts commonly feel compelled to address the relative importance of genetic and emotional bonds and the validity of "genetic rights."[15] This was the issue in the widely publicized dispute over "Baby Jessica" (see Chapter 4). Though popular sentiment strongly favored leaving the child with the DeBoers, who had raised her for two years, the legal circumstances of the case clearly favored the claims of her biological father. In a very similar case in June 1994, a judge in the Illinois Supreme Court ordered a three-and-a-half-year-old boy, "Baby Richard," removed from the home of his adoptive parents and returned to his biological father, Otakar Kirchner, who had never seen him. Unlike Baby Jessica, however, Baby Richard had been legally adopted before his biological father appealed for custody. Kirchner asserted that the mother of the boy told him the baby had died, so he did not learn the truth until after the adoption had gone through. The Illinois court's ruling revoking the adoption has been appealed, and Baby Richard remains with his adoptive parents in suburban Chicago. But his biological father has announced that he will "never give up" his efforts to obtain custody: "Adoptive parents can replace the kid. For biological parents, it's not so easy. Your own kid is your own kid."[16]

The importance of genetic bonds has also been debated when teenage children seek to "divorce" their biological parents. In 1992, twelve-year-old Gregory Kingsley went to a Florida court to end the parental rights of his genetic mother and to allow his foster family to adopt him. Gregory had been living with this foster family for two years, after years of abuse and neglect by his biological parents. In this case, the court considered the indisputable evidence that Gregory had been abused as overriding the biological link; it allowed the child to sever legal connections with his biological mother. The decision, however, was controversial, criticized as a "dangerous precedent" threatening "the preservation of the family." To critics of the decision, biological connection appeared more important than the social experiences of the child.[17] Following Gregory's success, fourteen-year-old Kimberly May—who had been switched at birth with another infant—sought to sever connections with her biological parents. She too won her case, though she later changed her mind and moved in with her genetic parents (see Chapter 4).

The concept of genetic rights remains contested, but privileging biological connections is attractive as a way to clarify ambiguities and resolve emotional disputes. Genetic ties can appear to simplify the options, defining difficult family dilemmas as molecular problems and permitting the resolution of disputes on the basis of scientific testing. Ideas about genetic predispositions have also successfully been incorporated into criminal law, where they are guiding decisions about responsibility and punishment for crime.

Punishing and Preventing Crime

The American legal system is built on the premise that responsibility, necessary to attribute culpability, rests on volition or free will (see Chapter 7). To be responsible in the legal sense, a person must be able to exercise choice freely and to control his or her behavior. This definition of responsibility underlies the insanity defense. Any biological condition, moreover, that seems to impair the ability to exercise choice or to control behavior can become the basis of a legal defense. Biological defenses have included such diverse conditions as the influence of sugar (the "Twinkie defense"), Pre-Menstrual Syndrome, postpartum depression, and the possession of an extra Y chromosome. Defense attorneys have used each of these conditions to argue that their clients acted involuntarily and were, therefore, not properly subject to punishment.[18] Indeed, a wide range of biological conditions appear in the legal literature as absolving a criminal defendant from blame.

In this context, the legal significance of genetic explanations lies in their bearing on free will. Those who defend themselves on genetic grounds, defining genetic predisposition to addiction or to violence as the cause of a criminal act, anticipate that this will be viewed by the court as a full defense—or, at the very least, that it will mitigate punishment.

The influence of assumptions about genetic predisposition on sentencing decisions can be seen by comparing two similar cases considered by the Supreme Court of California in 1990, *Baker v. State Bar*[19] and *In re Ewaniszyk*.[20] Both were disbarment proceedings against attorneys who admitted they had misappropriated client funds. Each attributed his crime to substance abuse, and both had apparently learned to control their problem. Yet Ewaniszyk was disbarred while Baker, who claimed a "genetic predisposition" to alcoholism, was merely placed on probation. Why should the genetic factor matter if both attorneys posed similar dangers to their clients and demonstrated equivalent capacity for rehabilitation? In effect, Baker's genetic endowment mitigated his punishment because the court assumed that predisposition precluded free will. Driven by his genes, Baker could not be held fully responsible for his actions.

As images of "criminal genes" or "alcohol genes" shape popular understanding, claims of genetic predisposition are likely to enjoy greater public and legal acceptance. If so, cases like *Baker* and *Ewaniszyk* raise a broader question: How will society adequately protect itself from those who constitute a genetic threat? Logically, the California court could have made the opposite choice, disbarring the genetically addicted Baker rather than Ewaniszyk: Since substance abuse was not identified as a part of Ewaniszyk's genetic constitution, perhaps he was *more* capable of complete reformation than Baker, whose genetic predisposition could doom him forever to battle his own nature. If this interpretation were to guide such decisions (as it did in Kerr's fictional society), genetic predisposition could be used to argue for *enhancing* rather than mitigating punishment.

The 1992 Hollywood film *Alien III* presented a science fiction extension of this dilemma. It featured a prison planet full of XYY males convicted of crimes and permanently deported, since they were believed to be incapable of rehabilitation. Their genetic predisposition to crime became, in this fantasy scenario, a justification for permanent abridgement of their freedom. In an episode of the television program "Law and Order," moreover, when a defense attorney invoked an XYY defense to mitigate the sentence of a teenager convicted of murder, the young man himself asked for a maximum sentence: He said that his situation was hopeless, because he was genetically predisposed to kill.[21]

Sentencing decisions are often based on predictions of dangerousness, that is, on the probability that the defendant will continue to constitute a threat to society. Most courts use social indicators like the nature of past behavior or the criminal's family context to make such predictions.[22] But the concept of genetic essentialism suggests the added importance of biological predisposition, an apparently permanent risk. Just as those acquitted of crimes by reason of insanity can be removed from society for longer periods of time than the maximum criminal sentence for their offenses, so those deemed genetically disordered could be permanently isolated. Though not willfully responsible for their actions, they may find their liberties circumscribed, for if violence reflects a genetic predisposition, rehabilitation may have minimal effect on the deterrence of future crime.

The past failures of rehabilitation schemes, rising costs of prison services, and the politicization of penal policy in the climate of skepticism about social welfare—all have contributed to a decline in the rehabilitation ideal.[23] The idea that those who violate laws can be transformed (by a model prison system) into productive, law-abiding citizens no longer guides the policy debate over prison management. In the criminology literature "selective incapacitation" (that is, removing the lawbreaker from society) has replaced rehabilitation as a goal. And criminal tendencies, whatever their original cause, appear to be fundamental and unchangeable. One psychologist, writing in a popular science magazine, announced that the criminal is "a different species entirely," adding that this had implications for rehabilitation. "There is nothing by which to rehabilitate a criminal," the psychologist said; "There is no earlier condition of being responsible to which to restore him."[24]

Molding the institutional uses of genetic essentialism in the 1990s are both concerns about the administration of criminal justice and growing public fear of crime. According to a National Academy of Sciences report, the United States is no more violent today than it has been at many times in the past. But the level of public fear has significantly increased, outpacing increases in the actual rates of violent crime.[25]

Perceptions of crisis can sanction suspension of rules and freedoms, if control comes to seem more important than civil rights. In a 1992 letter to the editor of *The Journal of the American Medical Association* (*JAMA*), for example, a physician insisted that "drastic times require drastic cures. . . . We cannot let such matters of conscience as the Second Amendment or preservation of personal freedoms stand in the way of public safety."[26] A writer for *Science Digest* wondered whether children who are suspected of being genetically prone to criminal behavior should be isolated or operated on, just as defective cars are recalled to the factory.[27] Travellers on American Airlines could read in the company's free magazine, *The American Way,* that "there are some who believe rape could be reduced greatly if we had a way to determine who was biologically predisposed to it and took preemptive action against them."[28] And a columnist in the *Washington Post* suggested that early intervention would be justified to stop those who are identified as genetically predisposed to violence. After all, he wrote, "It seems pointless to wait until the high risk prospects actually commit crimes before trying to do something to control them." He did, however, also recognize that genetic information about predisposition could not be considered the equivalent of an act of violence, that it would be "unconstitutional and unconscionable" to convict anyone on the basis of predisposition without evidence of an actual crime.[29]

This crisis mentality shaped the creation of the Bush administration's Violence Prevention Initiative. The Bush program grew out of the belief that crime must be approached as a public health problem and treated as if it were an epidemic. Preventing the spread of infection required the identification of "natural carriers" assumed to be "biologically prone" to criminal behavior. The program therefore focused on early detection of those at risk, especially in "epidemiological cachement areas" (the inner cities). It called for screening about 100,000 inner-city children to identify potential criminals. Those who appeared likely to cause problems would then be sent to special day camps where they would participate in remedial, behavioral modification programs to help them avoid their fate.

Responsibility for crime control remains located in agencies dealing with matters of public health: the Department of Health and Human Services and the Centers for Disease Control. Applied to crime control, the public health model suggests that, as in the case of epidemics, individual rights must be suspended to preserve public order. Reinforced by popular belief in the deterministic powers of the gene, genetic information could open opportunities for social control of unprecedented power as predispositions are employed to predict and avoid potential risk.

Predicting and Avoiding Risk

A cartoon portrays Ms. Tena, "reader-advisor," and her competition at the shop next door, Madame Rosa, "geneticist." Each promises to predict the future.[30] Crystal balls are generally consulted by individuals interested in predicting their own future, but the crystal ball of DNA analysis has increasingly become a resource not for individuals, but for the social institutions that shape their lives. Many institutions and organizations—schools, motor vehicle bureaus, immigration authorities, sports teams, organ transplant registries, adoption agencies, the military, even university tenure committees—have significant interest in predictive information. Pressured to maintain economic viability, efficiency, and accountability, such institutions need to anticipate future

contingencies and to avoid risks. Genetic tests can differentiate individuals on the basis of ostensibly natural categories
by identifying predispositions. Their results, applied by
institutions seeking to control costs, appear to resolve ambiguities while limiting the role of arbitrary interpretation.[31]
They can end up defining the competence of individuals,
controlling their access to social services, educational programs, insurance, or jobs.

Medical institutions, especially health insurers, face
economic pressures that encourage interest in predictive
information. Genetic information can help health care
providers conform to the reimbursement constraints of
third-party payers, control access to medical facilities, and

plan for future demands in a system increasingly preoccupied with cost containment and economic accountability. Insurance underwriters, favoring "preferred" customers who are expected to incur fewer medical expenses, welcome genetic information as a means to cut their costs. People with a genetic predisposition to a costly disease may be denied insurance coverage or expected to pay exorbitant rates.[32] Some health insurers have already tried to refuse coverage for children born with birth defects after the mothers, warned through prenatal testing, refused to abort.[33]

The threat of malpractice litigation also drives the interest in genetic information. The courts have recognized "wrongful birth" claims, awarding damages to parents of children born with genetic disorders that could have been identified in time for a fetus to be aborted. This encourages the use of genetic testing as a way to protect physicians against malpractice suits.

As the scope and sophistication of testing for genetic characteristics increases and their use becomes widely accepted, companies could justify routine testing of prospective employees for their predisposition to personality disorders and to traits that may affect their ability to work. Some corporate wellness programs offer genetic tests to employees in order to help them (and the company) anticipate future problems. Job application procedures could add genetic tests to the existing battery of pre-employment examinations, enabling employers to hire workers with genetic profiles that suggest they are likely to stay healthy and perform well.[34] Just as psychological tests are now used to predict potential productivity, honesty, and special skills, so biological tests could be used to anticipate future failure or success.

At present, genetic screening techniques are mainly used to identify and exclude workers with specific traits that may predispose them to illness from exposure to certain chemical agents in the workplace.[35] Justified in the first instance as a way to protect employee health, tests that identify vulnerable individuals can also be used to control compensation claims and avoid costly changes in the workplace environment, reducing a company's burden of responsibility to provide safe working conditions. It is the employee, excluded from the workplace, who assumes the burden and the blame for the risk.

Different corporations have stakes in genetic explanations for quite diverse reasons. Information about genetic predispositions can be used to deflect a company's legal responsibility for doing harm by blaming the damage on the victim. For example, when a behavioral modification clinic was sued following the suicide of a young man who had been attending its program, the clinic defended its procedures by arguing that the man was genetically predisposed to mental illness. It called in geneticists as expert witnesses to attest to the importance of genes as causes of suicidal tendencies.

The Ernest Gallo Clinic and Research Center, staffed by neuroscientists at the University of California at San Francisco and funded by the eighty-five-year-old wine mogul, has another stake in genetic explanations. Created in 1984, the center has worked to identify the biological causes of alcohol abuse. In 1993, Gallo Center scientists found what they believed to be a gene for alcoholism. They hypothesized that this gene produced a protein that "jams the signals" that would normally warn a person to stop drinking. Those born without this internal warning system, they suggested, might be prone to drink too much. Critics of the Gallo Center, which has received millions in research funding from the manufacturer of inexpensive wines such as Thunderbird, argue that a genetic explanation for alcoholism might be extremely useful to the liquor industry. A biological cure for alcoholism could "increase the potential market for their product," while also locating responsibility for alcoholism in an individual's DNA.[36]

Economic pressures on government agencies can also drive genetic testing. Take, for example, the Fragile-X testing program in Colorado. Fragile-X is a genetic disease that can, but does not always, cause mental impairment. To justify genetic testing for the disease, proponents of the Colorado program provided an estimate of the public cost of caring for Fragile-X patients. They concluded that the economic burden called for improved identification of carriers—"even mildly affected carriers of normal IQ"—so as to reduce the number of affected births. "The savings for the state would be tremendous."[37]

Similarly, on Long Island (in New York State), members of an organization seeking to reduce school taxes have campaigned against the school system's program of special education classes for learning-disabled children. Their argument is that learning disabilities are of genetic origin; therefore the responsibility falls to the medical system, not the schools.[38]

Genetic assumptions, focusing attention on the aberrant individual and away from the social context, also have strategic value for educational institutions facing demands for accountability and pressures to establish more rigorous standards for classifying students. During the 1960s, explanations of academic success or failure centered on environmental sources of behavioral and educational problems, calling attention to the influence of family deprivation or a child's socioeconomic situation. But in the 1980s, social explanations were gradually replaced by explanations drawing on the biological sciences. Learning disabilities and behavioral problems became defined as biological deficits, with problems located less in a student's social situation than in the biological structure of his or her brain.

These ideas have been reinforced by the Diagnostic and Statistical Manual of Mental Disorders (DSM) published by the American Psychiatric Association. Its classifications serve as important guides for educational counselling. During the 1980s, in the course of various revisions, the DSM shifted away from psychoanalytic assumptions toward an approach increasingly based on biology.[39] The 1980 edition had added the biological categories of "learning disability" and "attention deficit disorder," defining such problems as internal brain states of the student, independent of his or her social group. Later revisions further differentiated these categories, adding for example "sterotypy/habit disorders" and "expressive writing disorders." The increasingly refined classifications implied that such problems are developmental disturbances—organic disorders rather than simply the result of inattention or problematic classroom management.

Schools, having access to most children in the society, are traditionally responsible for assessing, categorizing, and channeling them toward future roles. Educators' interest in biological causes, therefore, has broad social impact, and their widespread acceptance of genetic essentialism legitimatizes the use of tests in the schools to define and measure what is normal or pathological in individuals, to assess their potential, and to predict their intrinsic limits.

The Blueprint of Destiny

The popularity of the Human Genome Project, with its almost weekly discovery of new genes and promises of new cures, encourages the institutional use of genetic information and, at the same time, discourages serious public scrutiny. According to the media and other sources of popular information, science provides objective, certain knowledge of genetic processes, which seem to predict, unfailingly, the future health of individuals. Yet these popular expectations of objectivity, predictability, and certainty may not be borne out when scientific studies are applied to specific social or institutional policies.

Objectivity is a norm in the practice of science, but the history of science demonstrates how cultural forces and institutional needs can shape the choice of research topics, the nature of scientific theories, and the representation of research results. Scientific fields are influenced by institutional agendas and defined to reflect the priorities and assumptions of given societies at particular times.[40] Historian Nikolas Rose, for example, has traced how the development of psychology as a credible scientific discourse in the early part of the century was driven less by curiosity about the human psyche than by the needs of the schools, the reformatories, the army, the factories, and the courts. Faced with rapid social change, these institutions sought to identify individual differences and to predict future pathologies in order to meet evolving demands. The field of psychology developed, therefore, as the "science . . . of individual differences, of their conceptualization and their measurement, . . . and of the prognosis of future conduct in terms of them."[41]

The recent interest in genetic perspectives stems not solely from dramatic advances in research, but also from the appeal of a scientific explanation that seems to justify social agendas. And our continued fascination with the criminal brain—currently reflected in the efforts to visualize the brain as a way to identify the biological basis of criminal tendencies—reflects the hope for simple, scientifically based solutions to the complex problem of understanding and managing crime. One criminologist, writing on the insanity defense, claimed that new genetic and imaging technologies will provide "direct scientific evidence of the defendant's brain . . . and we need not rely on verbal reports and questions such as 'Did the defendant know right from wrong?'"[42] Philip Kerr's Lombroso Program seems less fantastic in this climate of intellectual anticipation.

The promise of prediction is seductive. In recent years, growing perceptions of risk have encouraged the development of several fields of study—technology assessment, risk analysis, social forecasting, and even "futuronics"[43]—devoted to predicting and controlling risks. Actuarial reasoning has become a guiding principle, a way to anticipate future contingencies. The popular appeal of genetics—focusing on the "oracle of DNA," the "blueprint of destiny"—lies partly in its image as a predictive science: a means to uncover predispositions.

Predisposition, however, is a malleable concept that changes when exported from clinical genetics to social policy.[44] The scientific concept of genetic predisposition assumes the existence of a biological condition signaling that an individual may suffer a future disease or behavioral aberration. But predisposition in the clinical sense is a statistical risk calculation, not a prediction. A person "predisposed" to cancer, for example, may have biological qualities that heighten the odds that he or she will develop cancer, in the same way that driving many miles each day heightens one's odds of involvement in an automobile accident. But many variables influence whether a person will actually suffer from cancer. Terms such as "predisposed" or "at risk" are under-

Stuart Goldenberg

stood by scientists to mean that the individual is vulnerable to a disease that may *or may not* be expressed in the future. In the quest to identify genetic predispositions, however, the statistically driven concept of correlation is often reduced to "cause." And possible future states, calculated by statistical methods, are often defined as equivalent to current status. A genetic predisposition to alcoholism, for example, becomes an "alcohol gene." And an individual "at risk" may be regarded as deserving differential treatment long before it is known whether or not the risk will materialize.

People diagnosed as predisposed to a behavior or disease may find themselves treated as if their fate were certain, even when the relationship between genetic defects and their manifestation in actual behavior or illness is conditional and poorly understood. Take, for example, the middle-aged man who in 1992 discovered that he had a genetic predisposition to a particular neuromuscular condition. He experienced no obvious symptoms and, in particular, had never had an automobile accident or traffic violation in twenty years of driving. But when his insurance company learned (through his medical records) about the predisposition, it refused to renew his automobile insurance policy.[45]

Research in molecular biology is yielding powerful new information that may reduce ambiguity in some areas and define meaningful constraints in others. Institutions are understandably attracted to the predictions promised by a genetic "map" and to the certainties implied by genetic essentialism. But predictive information that helps institutions control uncertainty and contain costs can also have devastating effects on individuals.

Genetic information can extend discrimination to new categories of persons. Those who carry the traits for certain disorders can be reconceptualized as the "predisposed," as "persons at risk" whose potential condition differentiates them from the "normal" and labels them unsuited for normal opportunities.[46] If an employer, educator, or insurer can make the case that the "predicted" future status of their client matters, then discrimination—denial of opportunity for medical care, work, or education—can occur with impunity. Indeed, predictive genetic typing may create an underclass of individuals whose genes seem to have marked them for the nowhere track.

Some of the more discriminatory applications of genetic information are limited by legislation. The Americans With Disabilities Act, implemented in January 1992, limits pre-employment testing to the assessment of a person's actual ability to perform a job. The act will help to limit the abuse of tests that indicate genetic predisposition in a person with no symptoms. But the legislation does not preclude the use of "sound actuarial data" as a basis on which to limit health care benefits. The standard underwriting assumptions of the insurance industry are not affected. And while laws may curb specific institutional abuses, attitudes that support informal discrimination are harder to control through regulatory schemes.

The social value placed on privacy could also help to limit institutional abuses of genetic information. But a recent public opinion poll suggested that, to many Americans, genetic information does not appear to be "private." A 1992 survey by the March of Dimes Birth Defects Foundation found that most Americans believe genetic information is public property and that those with a right to information about a person's genetic characteristics include not only those family members who could be immediately affected, but insurers and employers as well.[47] The survey also found that 43 percent of Americans approved of using gene therapy to enhance the physical and behavioral traits of their children as well as for treatment of disease. The survey respondents clearly understood little about the details of genetics and the potential for harm.

In this context of public naïveté, popular assumptions about the powers of the gene make it more difficult to control institutional abuses of genetic information. Laws prohibiting discrimination may, in practice, prove less important than public acceptance, fed by ubiquitous images and appealing stories, of the essentialist premises underlying institutional decisions. So long as persons continue to be conceptualized as aggregates of physical attributes and as gene-transmitting agents, biology can be used as both a standard for opportunity and a justification for discrimination.

In the rush to apply the latest research, institutions may oversimplify the complex and poorly understood relationship between genetics and environment. And in the urgency to find solutions to social problems, they may compromise or obscure important values of equality, justice, and privacy. Genetic screening intended to assess intelligence, criminality, or predisposition to learning or behavioral problems can affect the quality of education, the functioning of the legal system, and the fairness of the work environment. Placing responsibility for social problems on the traits or predispositions of certain individuals can justify policies of discrimination or exclusion in the interest of enhancing efficiency or maintaining social control. Ultimately, the world view of genetic essentialism leads to policies that restrict the reproductive rights of individuals, for it suggests that order in a society depends on the genetic qualities of its population.

9

Genetic Futurism

A 1991 letter to the *New York Times* reiterated an old theme: human society, the writer warned, had taken a "Darwinian U-turn"—"humankind has done little to improve its own breed. ... Worse, our social programs encourage many less genetically endowed to breed."[1] A conservative writer in 1992 claimed that America's economic problems could be attributed to the "lack of selection pressure" in the United States.[2] And an editorial writer for the *Philadelphia Inquirer* proposed in 1990 that a new passive birth control technique should be used to eliminate poverty: Prevent the poor from having babies and you will eventually have no more poor.[3]

The social policy imperatives encoded in these contemporary pronouncements were, of course, known as eugenics in the late nineteenth and early twentieth centuries. Today, in response to widely publicized research in molecular genetics, many observers have explored the likelihood of a "new eugenics." They postulate a eugenics fueled by the latest techniques for manipulating DNA, by public concern about rising health care costs, and by social tensions over poverty, crime, and race. Critic of science Jeremy Rifkin anticipates a future in which multinational corporations will "manage production by controlling vast areas of the earth's commons—the land, sea . . . and the gene pool."[4] But historian Daniel Kevles expects that a new eugenics will *not* recur, blocked by the democratic nature of our social institutions, public awareness of the historical abuses of state-sponsored eugenics, and scientists' more sophisticated understanding of the limits of genetic intervention.[5]

Between these views are a wide range of other speculations. Sociologist Marqui Luisa Miringoff, for example, sees in this policy debate a shift from a rhetoric of social welfare to one of "genetic welfare." The former looks to social structure as the source of social problems; the latter looks to individuals, assuming that problems can be solved by eliminating those who caused them—by implication, those who are genetically flawed. Miringoff predicts "a drift toward a variety of modes of genetic determinism, the increasing delineation of our genetic rights and duties, and the widening effort to reduce the genetic burden."[6]

Other observers see the threat of a new eugenics as less direct. Sociologist Troy Duster argues that while the "front door to eugenics is closed," the "back door" has already been opened by contemporary medical care practices such as genetic counseling, the selective identification of genetic disease, and the conception of genetic health.[7] Historian Carl Degler suggests that new eugenics policies will not rely on compulsory state intervention, as they did at the height of the American eugenics movement, but on voluntary compliance based on perceived economic and social interests and a sense of eugenic responsibility.[8] Similarly, writer Richard Neuhaus believes that eugenics is returning with the "manufacture of

170

synthetic children, the fabrication of families, artificial sex, and new ways of using and terminating undesired human life." He sees the language of eugenics today as "the winsome one of progress, of reason, and, above all, compassion."[9]

From our perspective, the rise of a new eugenics—mediated not by state policy but by social and institutional pressure—is made more likely by certain ideas conveyed in popular culture. Stories of genetic essentialism and biological determinism facilitate public acceptance of the control of reproduction for the common good. The idea that human beings need to seize control of human evolution is expressed in popular sources by means of four related ideas. The first is that different rates of reproduction among groups (differing in class, race, ethnicity) threaten the quality of the human gene pool or the future health of the species. The second claims that there are "lives not worth living"[10]: people who simply should not be born. The third asserts that environmental and economic problems are a consequence of reproductive practices. And finally, some conceive the threat as severe enough to justify limitations on reproductive rights—such limits to be enforced, if necessary, by the powers of the state. We will show that, in many popular sources, the control of human reproduction appears as a solution to pressing economic and social problems, a way to avoid disaster, and a means of controlling the future.[11]

Differential Reproduction: "The Poor Breed More"

In 1991 CBS commentator Andy Rooney made the comment that "most people are born with equal intelligence but blacks have watered down their genes because the less intelligent ones are the ones that have the most children."[12] When the network temporarily suspended him, he had many supporters. An editorial in the *Washington Post* criticized the "sanctimonious cultural fascists of the left." The editorial writer associated the network's sensitivity to eugenic and racist language with a "wave of high-minded suppression that seems to be sweeping the country."[13]

Rooney's reference to watered-down genes drew on a form of evolutionary explanation that played an important role in earlier eugenic initiatives. The popular idea that the poor are uniquely fertile persists despite data showing that three-quarters of all families receiving welfare in the Aid to Families With Dependent Children program have only one or two children.[14] The image of black women producing large numbers of illegitimate children at state expense captures a nexus of social anxieties over crime, the deficit, and even America's ability to compete in global markets.[15]

Some observers blame America's economic decline on the lack of selection pressure in the United States—particularly the disproportionate reproduction of the poor. Comparing America's gene pool to that of Japan, they find that Japan has surpassed the United States in this arena, as in so many others. For example, a writer for the *Whole Earth Review*, David Kubiak, suggested in a 1990 report that samurai executions were a form of artificial selection and had produced a population ideally suited to the demands of contemporary corporate life. Kubiak said that the "the taste for organizational life" is genetic and, like most genetic traits, "randomly distributed." In Japan, independent types who abhorred hierarchy and authoritarianism had been decimated by the "joiner" types. As a consequence, the independents' "gene pool slowly began to bleed away." During the Edo period, the samurai killed those deemed dangerous "with the same impunity that a breeder culls his flocks of undesired traits." Some fifteen generations of the population were thereby "genetically pruned" of "assertive and egalitarian DNA." This carefully bred population, said Kubiak, explains the success of Japan today.[16]

Others have attributed Japan's success to its repression of women—specifically, its limitation of women to caregiving roles. Daniel Seligman, a prolific writer and an editor of *Fortune,* calls the American belief in equality and the importance of environment "pious baloney." The problems of American schools may partly account for American decline, but the real explanation, he says, is that the poor are reproducing faster than the rich. In a 1992 book, Seligman contended that IQ was responsible for economic status: "The rich are more intelligent than the middle class and the middle class are more intelligent than the poor." The United States is slipping, he wrote, because of above-average birth rates in the low-IQ population. He opposed "affirmative action for professional women" as a "dysgenic policy" that provides professional opportunities for highly educated women who then "invest more heavily in their careers and less heavily in raising families." He contrasted this with the situation in Japan, where "high-status women have just as many kids as poor women."[17] Intelligent women have the responsibility to reproduce, Seligman argued, for "America of 2050 would be a better place to live if between now and then the high-IQ population had more kids than the low-IQ population."[18]

Elite American women who fail to produce enough children were similarly taken to task by Richard Herrnstein, whose articles on race and IQ were controversial during the 1970s (see Chapter 6). In the 1990s Herrnstein has returned to prominence to warn of a future "confederacy of dunces" due to the falling birth rate among affluent, educated women. Genes for intelligence, he claimed in news articles and in a 1994 book with Charles Murray, are unevenly distributed among social classes because people marry within the social class into which they were born. Since the upper classes tend to have fewer children, a "shrinking of the genius pool" results, threatening the species with genetic decline.[19] And in 1987 Ben Wattenberg, a fellow of the American Enterprise Institute, attributed the "backsliding" of America to the neglect of reproductive responsibility among bright women.[20] These critics place the burden of responsibility on intelligent women whose professional interests appear to compete with their reproductive roles.

A similar idea shaped the plot of a 1985 "True Life" mystery novel—a story about a search for a kidnapped child. The child was the daughter of a Nobel laureate geneticist who had carefully selected a "perfect mate" so as to breed a child with "good genes." The mating experiment, said the geneticist, in a story that could have been written by Herrnstein and Murray, reflected his sense of social responsibility. "The gene pool is in serious trouble, you know. All the wrong people are reproducing. . . . All the brightest people, the ones who really ought to be reproducing, aren't. . . . If something isn't done, the next generation won't be fit to cope with the world."[21]

In these popular narratives, different rates of reproduction in different groups—rich and poor, Japanese and American, or intelligent and unintelligent—threaten social progress. Superior individuals have a responsibility to replicate their genes for the good of the species, the race, or the nation. But those who are flawed—who carry genes for disease or abnormality—have a different responsibility, to prevent perpetuation of "bad" genes.

Lives Not Worth Living

Antiabortionists are fond of citing a comment they attribute to geneticist Sir Francis Crick: "No newborn infant should be declared human until it has passed certain tests regarding its genetic endowment. . . . If it fails these tests, it forfeits the right to live."[22] This remark is an extreme expression of an often muted set of questions: Is it right to bring a "damaged child" into the world? Do carriers of genetic diseases have a duty to prevent their perpetuation? Are there some lives that are not worth living?

People with hereditary diseases and disabilities fear that increasing access to genetic information through prenatal screening, together with the increasing acceptability of selective abortion of "defective" fetuses, will devalue them and their experiences, leading to increased discrimination against those who are physically different.[23] They express their considerable discomfort at the terms commonly used in prenatal screening discussions, sensing "disdain, discomfort and ignorance toward disabled babies."[24] Deborah Kaplan of the World Institute on Disability fears that prenatal diagnosis "is a statement that disabled people shouldn't exist."[25] A special 1994 issue of the *Disability Rag*, called "Eugenics, Abortion, Prenatal Testing," warns its readers of a "blooming of eugenics" inspired by advances in science that will be used to "keep people with disabilities from entering the world."[26] And Laura Flynn, the executive director of the National Association of the Mentally Ill, worries that, in light of the stigma attached to those who suffer from mental illness, many people would wish to see genetics used to "eliminate" them. "What happens if we find a gene? Will there be pressure to limit procreation by people with mental disorders?"[27]

A public dispute in 1991 over the reproductive choices of a well-known television anchorwoman reinforced such concerns. Bree Walker Lampley has the hereditary condition called ectrodactyly, manifest in abnormal hands and feet. As a celebrity, she was subject to public scrutiny not only of her physical differences, but of the choice she and her husband made to become biological parents despite their 50 percent risk of bearing a child with the disability. In 1988, when she was childless and working at WCBS in New York City, her co-anchor Jim Jensen asked her on the air whether her parents would have considered an abortion if they had known about her condition, and whether she would consider an abortion if she herself became pregnant. A few months later, Lampley did become pregnant; her baby was born with ectrodactyly. Three years later, when Lampley was pregnant with her second child, a Los Angeles talk-radio host invited listeners to call in their opinions on Lampley's decision to bear a second child, after the first had been born with unattractive hands and feet.[28] The radio host, Jane Norris, expressed her own disgust with Lampley's decision. "It would be difficult for me to bring myself morally to cast my child forever to disfiguring hands ... I'm wondering about the social consciousness of it all."[29] She then asked her listeners, "Is it fair to pass along a genetically disfiguring disease to your child?"[30]

Some who responded said Lampley's decision was her own business; others said Lampley was "irresponsible," that her choice was a "horribly cruel thing," or that "I would rather not be alive than have a disease like that."[31] One caller, appalled by Norris's question, said that the host's tone of voice "kind of smacks of eugenics and selective breeding." Norris responded: "What's your problem, you have a problem talking about deformities?" To another skeptic, she said: "I think we have enough imperfections that we don't need to perpetuate." Lampley, offended by the broadcast, filed an FCC complaint, which was eventually dismissed.[32]

The incident shows how a physical difference can be constructed as a social offense. Lampley was interpreted by the radio talk-show host and her sympathetic listeners as a genetic threat—and, therefore, a suitable subject of public discussion. The anxiety over the "social consequence" of her decision reflected contemporary concerns about the disabled as a drain on increasingly limited social resources. Robert Bogdan's study of "freaks" has shown that the bodily traits that identified an individual as a "freak" in nineteenth-century sideshows depended on context, expectations, and cultural beliefs about human difference.[33] Once attributed to the "work of the devil" or "God's Curse," or viewed as a punishment for moral transgressions, disabilities today are often seen in terms of their cost implications for the state. When disabilities are understood as economic burdens for the larger society, whether or not to bear a potentially imperfect child becomes a social as well as individual choice. Lampley's transgression went further: As a public figure and a role model, she was expected to take socially responsible actions that would avoid perpetuating "bad genes."

The debate over Lampley was unusual in its focus on a clear physical disability and on a celebrity's personal life. More often popular stories about "bad genes" focus on violence or mental illness and invoke fictional characters who are used to explore the problems posed by this vision of reproductive responsibility. A revealing 1992 episode in the television series "Northern Exposure" portrayed Joel Fleishman, a doctor in a small Alaskan town, counseling a patient, Holling, who believed he was the direct descendent of Louis IV of France. Holling was worried that this ancestry, with its history of violence, meant that he was genetically prone to commit atrocities. He wanted to marry, but feared that he would pass on this violent nature to his children. The doctor agreed that Holling's genes were potentially destructive. "What you are talking about is a genetic Chernobyl," he said. Holling then told his girlfriend about his tainted ancestry, and they decided not to marry.[34]

Similarly, novelist Fay Weldon's *The Cloning of Joanna May*, (1989) featured Carl, a man who would not have children for genetic reasons. "Do you understand just what sort of inheritance I have? . . . What sort of bestial blood flows in my veins?" Carl's parents were mentally ill, so he had himself sterilized: "I would be the end of the line; that particular experiment of nature's. I chopped down the family tree."[35]

Some media stories, however, do emphasize the intrinsic value of life, even life that is genetically imperfect. The prime-time television series "Life Goes On," popular for several years until it was cancelled in 1993, was the story of a loving nuclear family in which one of the children had Down Syndrome. Actor Chris Burke, who has Down Syndrome himself, played the teenager. Not the typical Down teen, his character attended high school, dated, held a summer job, and dealt with normal adolescent fears and uncertainties as well as a strenuous course load.[36] His limitations played a role in the plots; yet the theme of the series was that his positive characteristics compensated for his weaknesses and that, despite his disability, he was a valuable and valued member of the family and of society. Nevertheless, when the boy's mother became pregnant again, she had amniocentesis. (The test showed a normal fetus, and the controversial issue of abortion was avoided.)

Television's popular "Star Trek: The Next Generation" conveyed a similar message about the value of disabled life. In an episode first aired in February 1992,[37] the *Enterprise* crew discovered a stellar fragment about to collide with a planet that was uninhabited except for a small group of humans living in a "biosphere." The biosphere inhabitants told the crew members that in only eight generations they had created a genetically perfect society. They had "controlled procreation to create people without flaws." Individuals were born destined for their careers—the leaders genetically suited for management, the scientists born to do research, and so on. Genetic defects were screened out before fertilization. "It was the wish of our founders that no one would have to suffer a life with disabilities." The *Enterprise*'s captain, Jean-Luc Picard, believed that this society had given away its humanity by breeding out the uncertainty, self-discovery, and acceptance of the unknown that make life worth living. He was inclined not to intervene in the collision; it was not disabled life that was without value, he said, but life perfected and engineered. In the end Geordi La Forge, the chief engineer of the *Enterprise*, saved the biosphere from destruction. La Forge had been born blind (but given "sight" by a special high-tech visor). After the cosmic rescue, he told a biosphere scientist: "Guess if I had been conceived on your world, I wouldn't be here now, would I? I'd have been terminated as a fertilized cell."

This "Star Trek" episode used a planetary disaster to affirm the value of disabled life. But the pervasive sense of urgency surrounding contemporary problems, in particular the state of planet Earth, has brought another perspective to the cultural view of the genetic future and its meaning for reproductive controls.

The Threat of Extinction

For some extremely varied groups—including neo-Nazis and ecological radicals—the biological condition of the human species is in crisis. The literature of these groups expresses a sense of emergency in the face of what they believe is imminent catastrophe. And though the problems they focus on are very different, they share a sense that the future of the planet will depend on the social and political control of reproduction. Their rhetoric places wider social and economic concerns in dramatic perspective, for their narratives are infused with values and beliefs that also appear, in less strident form, in mainstream popular culture.

The rhetoric of such right-wing groups as White Aryan Resistance, Aryan Nations, Skinheads, and the American Nazi Brotherhood draws on evolutionary arguments. "Nature may be ruthless in culling out the weak, the meek, the misfits and the degenerate," one pamphlet proclaims, but "this does not mean that Nature is cruel. On the contrary, in making

178

sure that the strongest, the healthiest, the most competent, and the best survive to procreate and bring in the next generation of a species, nature is carrying on its benevolent program of building a better species and a more orderly world."[38] White supremacists call for "white genetic procreation." One group newspaper, *Racial Loyalty*, advertises a mating service with the motto "white men and women, be fruitful and multiply! Do your part in helping to populate the world with your own kind!"[39] A sticker proclaims: "Earth's Most Endangered Species: THE WHITE RACE Help Preserve it."[40] The issue, says another, is "Survival, Expansion and Advancement of the White Race."[41] Some of this literature translates anti-Semitic articles promoting "extermination" from the 1940s German press: "In a healthy community . . . sickly elements are normally not allowed to reproduce."[42]

In a review of the hate movement, lawyer Mari Matsuda documented the increase in racist incidents during the 1980s, concluding that a "message of hatred and inferiority is conveyed on the street, in schoolyards, in popular culture, and in the propaganda of hate widely distributed in this country."[43] The number of hate groups jumped from 273 in 1990 to 346 in 1991, a 27 percent increase. Among these, neo-Nazi groups that explicitly sport Third Reich uniforms and symbols increased from 160 to 203.[44] Their propaganda has apparently become less offensive in a climate where some Americans have come to doubt that the Holocaust ever occurred.[45] In this context the gentrification of racism—the increasing number of racial and anti-Semitic incidents on American college campuses and in mainstream politics—has been a special source of alarm.

The striking, if brief, success of David Duke in the 1991 gubernatorial campaign in Louisiana and the presence of Pat Buchanan in the presidential primary indicate that extremist sentiments have a base of popular support. Duke, a neo-Nazi apologist, was president of the National Association for the Advancement of White People until 1990. He believes that race and genetics are the dominant features in human society, and from his childhood hobby of breeding rats he became convinced that "genes make a profound difference."[46] In 1988 he signed an article warning of a "Black Population Bomb . . . you and your actions over the next few decades will decide who will propagate and who will not, who will control and who will be controlled."[47]

Duke built his political campaign on the view that the strength of America lies in its "European descent"[48] and that this is being challenged. While no longer visible in national politics after the 1992 election, his message, conveyed in common references to "the rising welfare underclass," captured a widespread perception: "Poor black people are having babies on your tax dollars."[49]

For activists in American hate groups, reproduction is the forum within which racial tensions will be played out. Their rhetoric pits black against white in a struggle to populate the earth and therefore to control it. Controlling reproduction is central to their long-term political goals. For a very different sort of activist, reproduction is also central, though in a different way. To some members of the ecology movement it is the unrestrained reproduction of *all* human beings, rather than the differential reproduction of racial groups, that promises to destroy the human future. In the ecological debate, the combatants are the human species and the Earth. The catastrophe of overpopulation justifies social controls on reproduction with as much rhetorical force as does the catastrophe of evolutionary decline. And while racial purity is the concern primarily of limited fringe groups, the ecological health of the planet has become the focus of popular movements around the world.

The Population Crisis Committee promotes the need to reduce the world's "carrying capacity" to avoid "irreversible damage to the global environment" and threats to "global security."[50] Its director, Sharon Camp, warns of "environmental refugees" moving into urban areas where city services are "collapsing." The Committee targets illegal immigration to the United States on the grounds that: "All but one of the sending countries . . . are Third World nations with relatively high fertility and low incomes." It suggests that street children and crime and violence in urban areas are the result of demographic pressures from countries where populations

must be controlled and from poor immigrants who reproduce more than those who can care for themselves.[51]

Some ecologists use the language of survival, writing
about "eco-catastrophe," "lifeboat ethics," and populations
"out of control."[52] Paul Ehrlich, author of the best-selling
1968 book *The Population Bomb*, said again in 1990 that we
"have overloaded the planet's biological circuits and are
breeding ourselves into oblivion."[53] We are in "a state of
emergency," and population control is the "most urgent
priority" facing the world.[54]

Some extremists in the ecology movement are misan-
thropes, like the speaker at a Public Interest Law conference
who declared that "We, the human species, have become a
viral epidemic to the earth . . . the AIDS of the Earth."[55] They,
too, frame their agenda in terms of reproductive responsibility. Christopher Manes, author of *Greenrage*, has proclaimed
that humans have a social responsibility not to reproduce.[56]
The *Earth First Newsletter* announced: "Love your mother,
don't become one," and "Real Environmentalists don't have
kids." One article proposed oral sex as a means of birth
control, and another described AIDS as a form of population
control.[57] Subscribers write letters proposing limits on "the
basic right to breed." Their suggestions have included tax
penalties for children, bonuses paid for sterilization, prohibition of immigrant families with more than two children,
mandatory sterilization of mothers who give birth to drug- or
alcohol-damaged children, castration of convicted rapists, a
ban on funding for infertility research, mass vasectomies,
and a "new look at female infanticide."[58]

The sense that the human gene pool is in decline, that the
human future is threatened—intellectually, socially, and bio-
logically—is ripe for satire. Mocking the overload of fervor,
an environmental group has named itself "The Voluntary
Extinction Movement."[59] A speed metal group called
Megadeth has put out an album called "Count Down to
Extinction."[60] Rock musician David Byrne sings: "Monkey
Man, DNA, and Evolution, Slide on Down, Say goodbye to
Civilization . . . Evolution's going Backward." A yellow button
reads "Gene Police: You out of the pool." A slogan announces

"You are standing in the shallow end of the gene pool." The dysfunctional hero in a John Carey novel is "the ultimate victim of whatever's gone wrong with the gene pool."[61]

Comic books and science fiction build on a standard plot of extinction, but while the stories written during the 1960s and 70s reflected fear of radiation, contemporary plots commonly focus on genetic decline. In one of the popular "X-Men" comic books, an evil mutant named Apocalypse plotted to attack "normal" humanity: "In the name of Evolution we claim your worthless lives on behalf of our Lord and Master. For you and your ilk are the weak . . . you were born only to perish under the hooves of the Horsemen of Apocalypse."[62] Another comic called "Ex-Mutants" featured a campaign to "terminate genetic deviants."[63] And a Marvel comic called "Extinction Agenda" is about a country called Genosha that breeds mutants as slaves.[64] Often in comic books alien mutants come to save the human species (which, having lost its genetic diversity, is doomed). Through eugenic practices of interbreeding, the aliens infuse new vitality into the human gene pool. In "Marvel Super-Heroes," for example, the earth is populated by the Kree, a humanoid race resulting from interbreeding between Earth's residents and aliens after the human species had reached an "evolutionary dead end."[65]

The theme of improving the human stock and saving the species from extinction is a plot device in many novels. Olaf Stapledon's classic 1930 novel *Last and First Men* was republished in 1988 in anticipation of new reader interest. It featured a society in which the "first men" declined because their gene pool deteriorated. But an aggressive program of selective breeding resulted in "a vast diversification of stocks," which led to a new biological group blessed with hybrid vigor. Successive improved species, Stapledon wrote, were created using "a technique by which the actual hereditary factors in the germ could be manipulated."[66]

In David Brin's *The Uplift War* (1987), genetic manipulation led to an "uplift" of the species. Society agreed to breed human beings selectively for the sake of "wonderful species diversity," which was "the wellspring of all sentience."[67] And Nancy Kress, in her futuristic 1993 novel *Beggars in Spain*,

The Double-Edged Helix

Blue Eyes Control
Quality Control
Inborn Criminality Control
Compensation Claims Control
Insurance Exclusion Control
Selective Breeding Control

was less concerned about survival than the next step in human evolution. The biggest advance in creating a super-race, she wrote, was being implemented through genetic manipulation of the fetus to remove the need for sleep.[68]

Octavio Butler's 1987 novel *Dawn* traced events after genetic problems developed following a nuclear war. Aliens became "gene traders" and mated with humans from earth to produce a new species.[69] And Arthur C. Clarke's *The Garden of Rama* (1991) featured three humans aboard the spaceship *Rama* attempting to assure the survival of the species by having babies. To vary the gene pool, the two men took turns fathering the children.[70] "Island City," a 1994 two-hour prime-time television program, was about "unnatural selection . . . a thriller set in the future [2035] in which genetics has run amok." A "perpetual youth" vaccine backfires and causes genetic mutations, producing "hideous recessives," people who are ugly and violent. The unaffected escape to an underground city where they establish rigorous controls on reproduction in an effort to preserve the race. Among these controls is a "DNA implant" worn on the chest, color-coded to indicate genetic quality. Fraternization with persons having incompatible colors is socially forbidden.[71]

Whether in social satire, fringe propaganda, or science fiction, these narratives of contemporary popular culture suggest compelling social interest in controlling reproduction to secure the human future. This interest has led some to advocate intervention that would selectively limit the individual's reproductive rights.

Limiting Reproductive Rights

In 1994, a wave of policy-oriented books appeared about the importance of heredity, the "decline of intelligence in America," and the threats posed by the declining quality of the gene pool.[72] Most prominent was Herrnstein and Murray's *The Bell Curve*, a book that was widely reviewed, excerpted in magazines, and hailed as a major political event. In a dense text that featured hundreds of pages of data, Herrnstein and Murray proposed that IQ (genetically determined and differing in different races) explained the current state of American society. The authors asserted that all sociocultural barriers to personal advancement had now been removed, and therefore social success and high IQ are perfectly correlated: those at the top of society are also those with the highest IQs. African Americans, they said, are disproportionately at the bottom of the economic scale because they are biologically inferior. Herrnstein and Murray regard welfare and social services as a fertility policy that "subsidizes births among poor women . . . at the low end of the intelligence distribution." By supporting poor women and their children, they argue, the welfare state itself produces crime, poverty, and illegitimate children. Similarly, they expressed concern about the "dysgenic" behavior of elite women with high IQs who tend to have fewer children than low IQ women. This tendency, they said, is "exerting downward pressure on the distribution of cognitive ability in the United States." They feel that these problems, associated with "differential fertility," call for urgent policy measures which include abandoning remedial education because the results are not worth the cost (those with less intelligence can never become high achievers), abolishing social programs because they encourage poor women (presumed to have low IQs) to reproduce, and developing alternative social support to encourage elite women (presumed to have high IQs) to have more children. These measures would, they feel, improve the cognitive quality of the American population.[73]   "Success and failure in the American economy, and all that goes with it, are increasingly a matter of the genes that people inherit."

The Bell Curve attracted a mixed response, ranging from approval as a brave statement of the obvious to total rejection as an ideological tract. But this and other books of its kind have enjoyed wide coverage because they capture a set of beliefs increasingly pervasive in popular culture. A letter to Ann Landers frankly expressed a popular view of "the women and girls on Medicaid who are having three, four and five children. . . . Why should hardworking, responsible taxpayers have to subsidize the irresponsible, careless behavior of these three, four and five time repeaters?" This writer suggested that such women should have their tubes tied.[74]

The problems and costs associated with low birthweight babies, Fetal Alcohol Syndrome, and babies with AIDS have also prompted discussions about the rights of women to bear children if this conflicts with the interests of the larger society.[75] This issue crystallized in debates over the use of Norplant, a surgically implanted contraceptive that releases hormones preventing pregnancy for more than five years. The Food and Drug Administration approved Norplant on December 10, 1990, and only two days later the *Philadelphia Inquirer* editorialized: "Poverty and Norplant—can contraception reduce the underclass?" The editorial suggested that welfare mothers be offered incentives to use Norplant, because "the main reason more black children are living in poverty is that people having the most children are the ones least capable of supporting them."[76]

African American writers promptly attacked the editorial as racist, as an endorsement of genocide, and as "thinly disguised eugenics."[77] Anti-abortionists became unlikely allies when a syndicated reporter wrote that support of Norplant for this purpose, like support of abortion, was "rooted in economics or matters of convenience." It pointed the way to forcing sterilization of those perceived as a burden to society, the reporter said, recalling Nazi policies and quoting a German doctor who had warned that "the biological heritage of the German people is menaced" by families with "unsound heritage."[78] *Newsweek*, however, suggested that the *Inquirer's* editor "was on to something . . . The old answers have mostly failed. After the shouting stops, the problem will remain. It's too important to become taboo."[79] Then a radio talk-show host speculated that Norplant could be a solution to teenage pregnancy and suggested that all women reaching puberty be required to use it. A month later, a judge in Visalia, California, ordered a welfare mother and child abuser to use Norplant as an alternative to a longer jail sentence. The judge argued that the "compelling state interest in the protection of the children of the state supersedes this particular individual's right to procreate."[80]

As a form of temporary sterilization, Norplant appeals to many women as a contraceptive of choice. Unwanted pregnancies continue to be a serious problem for women of all classes, and Norplant ranks as one of the most dependable methods of avoiding such pregnancies. It is an extremely effective, passive form of birth control that need not be remembered, applied, or changed for five years. But the very qualities that make it attractive as a contraceptive of choice also make it a politically useful means of controlling reproduction for the public good. In a *Los Angeles Times* poll in 1991, 61 percent of the respondents approved of requiring Norplant for drug-abusing women of child-bearing age.[81] During 1991 and 1992, thirteen state legislatures introduced bills that proposed offering Norplant to women on welfare. All these bills failed to pass, though by narrow margins. Most were built on financial incentives. Kansas, for example, proposed that the women choosing to participate receive $500 initially and $50 annually as long as the contraceptive remained implanted. The purpose of the program, said its advocate, was to "save the taxpayers millions of their hard earned-dollars."[82]

Reviewing the Norplant response as part of a wider tendency toward forced contraception, lawyers Julie Mertus and Simon Heller see "a resurrection of old-fashioned eugenics: plans designed to improve society by ensuring that certain state-undesirables, namely low-income women and men of color, do not reproduce."[83]

The Norplant proposals had more to do with controlling the cost of public assistance to the poor than with the genes of the poor, though eugenic implications were apparent. The association became explicit in proposals by extremists. In an overtly eugenic book called *Life Child: The Case for Licensing Parents* (1992), Randall Fasnacht of the Life Force Institute proposed that Norplant be used to control fertility so that only those able to meet certain standards could reproduce.[84] These standards included "recognition of genetic and inherited personality characteristics." The weak and unhealthy, wrote Fasnacht, should not have children: "Their gene pool will not be extended beyond the current generation. The strong will survive; the weak will wither." He supported these views by arguing that the birth of children will have to be restricted or the human population will overwhelm the earth's resources—thus drawing on the ecological concerns of the population control movement to suggest that eugenics is both necessary and legitimate.

The Eugenics of Individual Choice

In the 1950s the influential American geneticist and Nobelist Herman J. Muller suggested that "purposive control" of reproduction should be carried out "not by means of decrees and orders from authorities, but through the freely exercised volition of the individuals concerned, guided by their recognition of the situation, and motivated by their own desire to contribute to human benefit in the ways most effective for them."[85] Muller later joined Robert K. Graham, the wealthy inventor of shatterproof plastic lenses, in founding the

Repository for Germinal Choice, a sperm bank that offered DNA from "superlative individuals," "the genes of men whose genetic inheritance seems exceptionally favorable."[86] In the 1990s, Muller's vision of reproductive responsibility and technological control has in some respects become a reality.

Sperm banks are growing in social acceptability.[87] In 1980, when Graham's sperm bank was the subject of significant media attention, the idea seemed shocking to readers of *Glamour* magazine. Forty-six percent thought the Nobel Sperm Bank was outrageous, 73 percent thought it dehumanizing, 11 percent, racially offensive; and 86 percent said they would not like to be inseminated with Nobel sperm.[88] The same year *Time* speculated whether brighter was better,[89] and *U.S. News and World Report* wondered if a "genetic elite" was taking shape: Was it a revival of Nazism?[90] *Parents* magazine also compared the effort to Nazism.[91] The conservative *National Review*, however, responded favorably, describing the megabrained scientists and noting that high-IQ people are, on average, healthier and stronger. "Essentially we tax disproportionally the able. . . . At the same time we pay indigents to produce more children."[92]

By 1992, there were more than 100 sperm banks in the United States. Graham continues to run the Nobel Sperm Bank, though only three Nobelists have donated their sperm. He advertises for sperm in the newsletter of Mensa, the high-IQ group that has founded a Eugenics Special Interest Group "to provide a communications network for all people committed to enhancing genetic quality."[93] There have now been 156 children produced with sperm from this elite group, according to Graham, and some of these have become popular interview subjects.[94] In 1984 Paul Smith, an employee of the Nobel Sperm Bank, left to start another bank, called "Heredity Choice," in Pasadena, California.[95] The names used by these reproductive service firms— "Select Embryos"; "Quality Embryo Transfer Company"[96]—project an image of the child as a genetic commodity.

At the same time, a wide range of biomedical technologies for the treatment of infertility have opened new possibilities for the selection of "better babies." The ability to implant embryos through In Vitro Fertilization (IVF) has led to the establishment of more than 250 IVF programs in the United States. It has also created a market for donor eggs, and advertisements offer from $2,000 to $5,000 for ova donation. Male medical students have been sperm donors for years. Now female students are selling their oocytes to pay their tuition (though the procedures, requiring hormonal treatment for super-ovulation, can be painful and risky).

Meanwhile new screening technologies allow preimplantation diagnosis of genetic and chromosomal abnormalities in embryos, which can now be tested for disease or for sex, then selectively implanted. Some IVF centers offer preimplantation diagnosis to couples with a familial history of a genetic problem, and the diagnostic repertoire can be expected to expand as scientists identify genes that predispose individuals to a growing number of conditions. Embryologists are also working on ways to manipulate the genetic makeup of gametes (sperm and egg) and zygotes (fertilized eggs) through nuclear transplantation. Using this technique, the healthy nucleus of a couple's zygote can be transferred into a fertilized donor egg from which the original nucleus has been removed, thereby circumventing defects in cytoplasmic DNA (inherited only from the maternal side and contained in the cell's cytoplasm instead of its nucleus).

As genetic tests are developed, they are used to identify carriers of genetic diseases. While justified as a way to enhance individual choice, they are also driven by economics. And genetic tests raise difficult questions about the meaning of "voluntary" controls of reproduction in the context of social and economic pressures.

In her 1984 short story "Pursuit of Excellence," Rena Yount explored the idea of voluntary eugenics. She portrayed a future eugenics mediated not by state coercion but by desperate parental love. In her society, wealthy parents bought genetic engineering services so that their children would possess spectacular physical beauty and intelligence. The bioengineered—the offspring of the rich—ran society, held all the top positions, made the most money, and solved the important scientific and technological problems. Yount's story focused on the anguish of a mother who could not afford genetic services for her first child, but who struggled to pull together enough money to buy intelligence and beauty for her second. In the end, her struggles tore the family apart. The story presented a critical perspective on both parental anxiety ("I want to do the best for my children") and genetic responsibility ("Make better people . . . and they will put the rest of the world in order. My daughter will be one of them.")[97]

Popular narratives about genetic responsibility commonly frame control of reproduction in terms of personal choice. These stories, based on the hopes and expectations of the American public, suggest an expansion of parental responsibilities. Those who plan to bear a child are expected to consider not only the emotional cost to the child (of a disability), but also the social and economic burdens on the state posed by less-than-perfect offspring. This interpretation expects welfare mothers to stop having babies because they recognize the cost to taxpayers; prosperous and successful women to have more babies because they recognize that society needs intelligent children; those with heritable disabilities to choose voluntarily not to bear children because they do not want to burden either the child or the society with the costs of abnormality; and all parents suspecting a "genetic Chernobyl" to choose not to reproduce. But all such decisions are to be voluntary, a form of preventive behavior based on personal recognition of social obligations.

Such narratives of responsibility, however, are politically naïve, as Yount's story suggests. Personal choices are socially mediated, influenced by cultural forces and economic realities. Individual decisions about reproduction reflect the assimilation of beliefs about desirable behavior and pressures from family and friends. Moreover, the privatization of reproductive decisions in the entrepreneurial culture of America leads to their commodification—as they did in Yount's world of purchased genetic talent, or as they do today in the market for donor gametes—and this is hardly likely to safeguard against abuse. The "voluntary" actions of individuals are constrained by economic circumstances and available opportunities. Few people have the resources to act in ways that drastically violate prevailing values; few can assume substantial risks in the absence of social support. As Hermann Muller long ago observed, social mores and collective values shape individual choices in ways that can fulfill eugenic ends, even in the absence of coercive public policies.

Extremist positions—the promotion of "positive eugenics" by conservative policy analysts and the "negative eugenics" espoused by neo-Nazi groups—remain marginal. Even the more modest proposals for state incentives to use Norplant have mostly failed to become official policy or law. They confront the American cultural belief that procreation is a human right and a private rather than a social choice.

Less marginal are the ideas about genetic responsibility that appear in the popular media. Eugenics in contemporary culture is less a state ideology than a set of ideals about a perfected and "healthy" human future. Commonly held beliefs about the powers of the gene and the importance of heredity facilitate eugenic practices even in the absence of direct political control of reproduction, for eugenics is not simply gross coercion of individuals by the state; even in Nazi Germany, individual "choice" played a role in the maintenance of a highly oppressive state policy.[98] Rather, it can be productively understood as a constellation of beliefs about the importance of genetics in shaping human health and behavior, the nature of worthwhile life, the interests of society, and, especially, the meaning of reproductive responsibility. These beliefs—conveyed through the stories told by popular culture—draw on the assumption that our social, political, and economic future will depend on controlling the genetic constitution of the species—the so-called human gene pool.

10

The
Supergene

In her presidential address to the Behavioral Genetics Association in 1987, Sandra Scarr recalled the hostility she and her mentor Arthur Jensen faced in the 1960s and 1970s for their attention to the genetic basis of IQ. With no apparent irony, Scarr compared those who questioned her research to Nazis and cast herself as a "freedom fighter" who had to overcome the objections of mainstream psychology in her efforts to show that IQ was primarily genetic and little influenced by training or environment.[1] Behavioral geneticists, she announced, had been "out on a limb" until about 1985, when "the tree trunk jumped under the limb." Now, she claimed, there was no longer any dispute among responsible scientists or the public about the strict genetic basis of human behavior.[2]

While her claims that the issue is resolved are insupportable, her assessment is correct in one respect. In the 1980s and, increasingly, the 1990s, genetic explanations have attracted many new followers. In a diverse array of popular sources, the gene has become a supergene, an almost supernatural entity that has the power to define identity, determine human affairs, dictate human relationships, and explain social problems. In this construct, human beings in all their complexity are seen as products of a molecular text. And this text appears in popular culture as the secular equivalent of a soul—the immortal site of the true self and the determiner of fate.

In the preceding pages we have documented the diverse powers of the gene, its sacralization, its colonization of new territories, and its politicization in narratives of social control and human difference. We have explored the role of DNA in the public imagination and its alleged relevance to social problems of crime and alcoholism, to the state of the American family, and to the changing dynamics of race and gender. We have described how the gene has become an essential entity that defines the individual and determines the human future. And we have shown that the gene in popular culture, although it derives some of its power from the prestige of science, is not limited by scientific data. As a cultural icon, its meanings mirror public expectations, social tensions, and political agendas.

What is the appeal of genetic essentialism in contemporary American society? In part, the power of this concept reflects the close relationship between prevailing theories of nature and cultural conceptions of social order. Perceptions of the natural order have often reproduced, and later justified, social arrangements. Conceptions of nature, as David Bloor put it, serve as "a code for talking about society, a language in which justifications and challenges can be expressed."[3] The claims of contemporary molecular genetics have recast long-standing beliefs about the importance of heredity in defining kinship relationships, individual behavior, and group characteristics. The "germplasm" or "blood" of the early twentieth century has become DNA—and as a highly precise molecular text, the hereditary material has gained new status without losing older connotations.

The gene is a malleable concept, and its plasticity also plays an important role in its widespread and frequent appropriation. Feminists invoke genetics in debates about motherhood and reproduction; conservatives in their efforts to promote the importance of family values; the disabled in their battles against stigmatization; antiabortionists in their arguments about when life begins; parents in their legal claims for custody of children; African Americans in expressing fears of genocide; homosexuals in their claims for civil rights.

Just as the idea of genetic essentialism can serve many different social agendas, so it intersects with important American values. Genetic explanations of behavior and disease appear to locate social problems within the individual rather than in society, conforming to the ideology of individualism. They also provide the equivalent of moral redemption or absolution, exonerating individuals by attributing acts that violate the social contract to the DNA, an independent force beyond the influence of volition. And genetic explanations appear to provide a rational, neutral justification of existing social categories. Conforming to the value placed on both personal responsibility and individual exoneration, consistent with the faith in scientific and technological progress, genetic explanations have considerable rhetorical force.

They are thus a convenient way to address troubling social issues: the threats implied by the changing roles of women, the perceived decline of the family, the problems of crime, the changes in the ethnic and racial structure of American society, and the failure of social welfare programs. The status of the gene—as a deterministic agent, a blueprint, a basis for social relations, and a source of good and evil—promises a reassuring certainty, order, predictability, and control.

The new molecular genetics is also appealing for its promise of medical "breakthroughs" and wonder therapies—of biological rather than behavioral or environmental controls of disease. Genetics, claimed a journalist in a 1992 article, is "the medical story of the century, [which] will dramatically cure cancer, heart disease, aging, and much more."[4] But consider also the implications for individual, professional, and social control. In popular narratives, controlling the body depends on understanding and manipulating DNA. Health depends on biological intervention rather than diet, exercise, emotional resiliency, or other environmental or behavioral attributes. The future of medicine seems to lie in more aggressive biological manipulation, rather than in social intervention to change behaviors that promote disease. Increased authority and power are therefore vested in scientists and physicians, who become the managers of the medicalized society.[5]

At the same time, the individual—formerly responsible as the source of social problems—is helpless before the powers of DNA. In the 1990s, the popular rhetoric of DNA is as much about loss of control and acceptance of biological fate as it is about cure and the control of our future. Traits that are "genetic" appear as immutable, deeply resistant to change initiated through individual action or external intervention. This clearly bears on both personal choices and social policies. A new kind of advice book (known informally as the "don't help yourself" book) has emerged in the 1990s, presenting the individual as profoundly limited by genetic destiny. Unlike the traditional self-help book, which encourages the reader to overcome personal weaknesses in pursuit of a higher goal, these texts advise the reader to accept personal limitations, for his or her own good. As a guide to personal behavior, moreover, the idea of genetic destiny also suggests that we can "eat, drink and be merry," forgetting self-denying quests for fitness and health, since "certain things are genetic and inalterable."[6]

On a larger scale, as a guide to social policy the idea of the passive and helpless individual seems to suggest the limits of human agency and intervention. The ideology of genetic essentialism encourages submission to nature and to constraints on the possibilities for social change.

This rhetoric—this conceit of a DNA destiny—is neither a necessary nor coherent way of understanding what it means for human culture that we are biological organisms. In a trivial sense every human trait is biological; whatever our other qualities, we are indisputably bodies. At the same time every human trait is environmental, for if we are not (as embryo, infant, child, adult) nurtured in a proper environment (with sufficient air, reasonable temperature, and adequate food) we do not become anything at all. The fact that we are particular biological organisms dictates a great deal about our culture and social behavior, although not necessarily in the ways generally expected. Our biological status has a profound impact on our most fundamental conceptions of morality and civic order: If we were predatory insects who consumed our young, would not our literature, law, and religion be very different?

It is also important to recognize that many traits that are clearly genetic are also malleable, influenced by culture and environment. Height is both genetically determined and dependent on nutrition. Similarly, such common conditions as myopia and allergic rhinitis are genetic, but can be corrected with eyeglasses and behavior modification (avoidance of allergens). We routinely manipulate genetic traits, changing them for fashion (hair color), health (crooked teeth), or social success (facial features).

Our purpose in this book has not been to argue that biology is irrelevant to culture—such a claim would draw a dichotomy as problematic as that of genetic essentialism. Rather, we are suggesting that genetic essentialism, for all its grandiose claims, is a narrow way of understanding the cultural meaning of the body. By elevating DNA and granting it extraordinary powers of agency and control, genetic essentialism erases complexity and ambiguity. Problems and opportunities both disappear behind the double helix that has loomed out of proportion in the social imagination.

The forces that encourage this ideology, promoting and legitimating its use, will also shape the future implications of genomics research. In a 1993 editorial, *Nature* urged scientific temperance in the promotion of this research, cautioning that the "triumphalism" surrounding genetic discoveries was leading to a "growing belief in a kind of genetic predestination." For the editors of *Nature*, the problem raised by this belief was one of public confidence: a too-zealous promotion of genomics, they warned, might lead to "disbelief, distrust, and resentment."[7] They called geneticists to task for raising public expectations too high.

We suggest a rather different construction of the problem. The findings of scientific genetics—about human behavior, disease, personality and intelligence—have become a popular resource precisely because they conform to and complement existing cultural beliefs about identity, family, gender, and race. The promises made by scientists reflect these beliefs. Such promises express the desire for prediction, the need for social boundaries, and the hope for control of the human future. At the same time, scientists' claims about the powers of the gene meet many social needs and expectations. Whether or not such claims are sustained in fact may be irrelevant; their public appeal and popular appropriation reflect their social, not their scientific, power.

The danger, then, is not that inflated promises threaten to backfire on the scientific community, but that such promises will long outlive their scientific utility. Designed to appeal to popular interests, they quickly acquire a life of their own. With its emphasis on the natural origins of human difference, genetic essentialism can threaten marginal groups; with its focus on individual pathology, it seems to absolve society of responsibility for social problems; and with its emphasis on reproductive controls, it opens the door to oppressive state or social practices. Today the futuristic fantasies of biological management encoded in genetic essentialism take on a sinister cast, since the new biomedical technologies of the fertility clinic or the doctor's office can so readily make them come true.

While many scientists are aware of these possibilities, they see the value of genomics research in its promise for a better future—a future free of devastating disability and disease. Geneticist James Crow, in a 1988 essay, described his hopes for a "compassionate eugenics," wisely administered, protective of individual rights.[8] James Watson, in a 1993 interview, foresaw the "humane" use of genetic information, which he expected would expand individual choice and enhance personal control.[9]

Yet our study of popular culture suggests another construction of the future impact of genomics research. We find that, for the individual, greater access to genetic information may have the effect of *limiting* options as these data intersect with institutional prerogatives, economic constraints, and, especially, public attitudes toward those who make dysgenic (and therefore costly) reproductive choices.

The narratives of mass culture give shape to what is seen in the world. They define what seems to be a problem, and what promises solutions; what we take for granted, and what we question. In the 1990s, these narratives present the gene as robust and the environment as irrelevant; they devalue emotional bonds and elevate genetic ties; they promote biological solutions and debunk social interventions. Such notions are already affecting the institutional and social uses of genetic information as popular images encourage consumers to accept policies and decisions grounded in assumptions about genetic essentialism.

As the science of genetics has moved from the laboratory to mass culture, from professional journals to the television screen, the gene has been transformed. Instead of a piece of hereditary information, it has become the key to human relationships and the basis of family cohesion. Instead of a string of purines and pyramidines, it has become the essence of identity and the source of social difference. Instead of an important molecule, it has become the secular equivalent of the human soul. Narratives of genetic essentialism are omnipresent in popular culture, here explaining evil and predicting destiny, there justifying institutional decisions. They reverberate in public debates about sexuality and race, in court decisions about child custody and criminal responsibility, and in ruminations about the meaning of life. The powers of the gene in popular culture—expansive, malleable, sometimes fantastic—derive not from evolutionary forces or biological mandates, but rather from social or political expectations. Infused with cultural meanings, the gene has become a resource that is too readily appropriated, too seldom criticized, and too frequently misused in the service of narrow or socially destructive ends.

Notes

PREFACE

1. M. Susan Lindee, *Suffering Made Real: American Science and the Survivors at Hiroshima* (Chicago: University of Chicago Press, 1994); Dorothy Nelkin and Laurence Tancredi, *Dangerous Diagnostics: The Social Power of Biological Information*, 2nd ed. (Chicago: University of Chicago Press, 1994); Dorothy Nelkin, *Selling Science: How the Press Covers Science and Technology*, 2nd ed. (New York: W. H. Freeman, 1995).

2. The cultural studies literature often differentiates between high and low culture. We find similar genetic images in a broad range of cultural products from contemporary art to the daily soaps.

3. For analyses of these debates and empirical studies of audience reception, see *Poetics*, 21, 1992, a special issue on the relationship of media and audience. Also Andrea LiPress, "The Sociology of Cultural Reception," in Diana Crane (ed.), *The Sociology of Culture: Emerging Theoretical Perspectives* (Oxford: Basil Blackwell, 1994).

CHAPTER 1

1. Advertisement, BMW of North America Inc., 1983.

2. John S. Long, "How Genes Shape Personality," *U.S. News and World Report,* 13 April 1987, 60–66.

3. Cartoon by R. Chast, *Health,* July/August 1991, 29.

4. Mervyn Rothstein, "From Cartoons to a Play about Racists in the 60s," *New York Times,* 14 August 1991.

5. Henry Howe and John Lynn, "Gene talk in sociobiology," in Stephen Fuller and James Collier, eds., *Social Epistemology* 6:2 (April-June 1992), 109–164.

6. "The Secret of Life" is the name of an eight-hour "NOVA" series, directed by Graham Chedd and aired on public television on 26–30 September 1993. The phrase is widely used in descriptions of DNA.

7. The gene is the fundamental unit of heredity. Each gene is arranged in tandem along a particular chromosome. A chromosome, the microscopic nuclear structure that contains the linear array of genes, is composed of proteins and deoxyribonucleic acid (DNA), the "double helix" molecule that encodes genetic information. Each gene generates, as the "readout" of its specific DNA sequence, a particular protein—its functional product in building the cell or organism. The 24 chromosomes in the human genome contain about 100,000 genes.

8. This term is used by Sarah Franklin in "Essentialism, Which Essentialism? Some Implications of Reproductive and Genetic Technoscience," in John Dececco and John Elia, eds., *Issues in Biological Essentialism versus Social Construction in Gay and Lesbian Identities* (London: Harrington Park Press, 1993), 27–39. She defines genetic essentialism as "a scientific discourse...with the potential to establish social categories based on an essential truth about the body" (34).

9. Joanne Finkelstein, *The Fashioned Self* (Philadelphia: Temple University Press, 1991). See particularly 177–193.

10. Anthropologists describe cultural differences in bodily skills that have far less to do with inherent biological limits than with social expectations. They find that bodily and mental capacities are shaped by social organization. They depend in great measure on and vary with social beliefs, practices, and techniques. Paul Hirst and Penny Woolley, *Social Relations and Human Attributes* (London: Tavistock, 1982), Chapter 2.

11. On the gene and its changing meaning, see Elof Axel Carlson, *The Gene: A Critical History* (Ames: Iowa State University Press, 1989), 23–38, 124–130, 166–173, and 259–271. To quote Carlson, "The gene has been considered to be an undefined unit, a unit-character, a unit factor, a factor, an abstract point on a recombination map, a three-dimensional segment of an anaphase chromosome, a linear segment of an interphase chromosome, a sac of genomeres, a series of linear subgenes, a spherical unit defined by a target theory, a dynamic functional quantity of one specific unit, a pseudoallele, a specific chromosome segment subject to position effect, a rearrangement within a continuous chromosome molecule, a cistron within which fine structure can be demonstrated, and a linear segment of nucleic acid specifying a structural or regulatory product. Are these concepts identical?... For some of these problems, the findings from different organisms are contradictory; for others, the agreements [between organisms] may be analogous rather than a reflection of identical genetic organization" (259). See also L. C. Dunn, *A Short History of Genetics: The Development of Some Main Lines of Thought 1864–1939* (1965; reprint, Ames: Iowa State University Press, 1991), 33–49 and 175–191; also, James D. Watson's treatment of the complexities surrounding the concept of the gene in his *Molecular Biology of the Gene,* 2nd ed. (Menlo Park, CA: W. A. Benjamin, Inc., 1970), 230–254 and 435–466. Watson notes that "even now it is often hard to identify the protein product of a given gene" (240) and "now . . . we realize that the rate of synthesis of a protein is itself partially under internal genetic control and partially determined by the external chemical environ-

ment" (435). Our point is that this is a complicated concept with a long, contentious history.

12. The corn geneticist and Nobelist Barbara McClintock once began a presentation at Cold Spring Harbor that captured the complexities of the gene in the molecular age. She proclaimed that "with the tools and knowledge, I could turn a developing snail's egg into an elephant. It is not so much a matter of chemicals, because snails and elephants do not differ that much; it is matter of timing the action of genes." Cited in Bruce Wallace's colorful reconstruction of the history of the gene *The Search for the Gene* (Ithaca and London: Cornell University Press, 1992), 176.

13. The status of genetic disease and genetic therapy is reviewed in *Science*, 8 May 1992. In that issue, see Daniel E. Koshland, "Molecular Advances in Disease," 717; F. S. Collins, "Cystic Fibrosis: Molecular Biology and Therapeutic Implications," 774–779; K. S. Kosik, "Alzheimer's Disease: A Cell Biological Perspective," 780–783; C. T. Caskey et al., "Triplet Repeat Mutations in Human Disease," 784–788; D. H. MacLennan and M. S. Phillips, "Malignant Hyperthermia," 789–793; E. Beutler, "Gaucher Disease: New Molecular Approaches to Diagnosis and Treatment," 794–798; E. H. Epstein, Jr., "Molecular Genetics of Epidermolysis Bullosa," 799–803; P. Humphries et al., "On the Molecular Genetics of Retinitis Pigmentosa," 804–807; and W. F. Anderson, "Human Gene Therapy," 808–813.

14. Neil Holtzman, *Proceed with Caution* (Baltimore: Johns Hopkins University Press, 1989), 88–105.

15. See the discussion of rhetorical strategies by Jeremy Green, "Media Sensationalism and Science: The Case of the Criminal Chromosome," in Terry Shinn and Richard Whitley, eds., *Expository Science*, Sociology of the Sciences Yearbook 9 (1985), 139–161.

16. See e.g. Walter Gilbert, "Current State of the H.G.I.," Harvard University Dibner Center Lecture, 15 June 1990.

17. Leon Jaroff, "The Gene Hunt," *Time,* 20 March 1989, 62–67.

18. Lois Wingerson, "Searching for Depression Genes," *Discover,* February 1982, 60–64.

19. Fergus M. Clydesdale, "Present and Future of Food Science and Technology in Industrialized Countries," *Food Technology,* September 1989, 134–146.

20. Daniel Koshland, "Elephants, Monstrosities and the Law," *Science* 255 (4 February 1992), 777.

21. Bruce Wallace, *The Search for the Gene* (Ithaca: Cornell University Press, 1992), 199.

22. Lucy Fellows, cited in John Noble Wilford, "Discovering the Old World of Maps," *New York Times,* 9 October 1992. See also Dennis Wood, *The Power of Maps* (New York: Guilford Press, 1992). The geographer Mark Monmonier has observed that "a good map tells a multitude of little white lies. It suppresses truth to help the user see what needs to be seen": *How to Lie with Maps* (Chicago: University of Chicago Press, 1991), 199.

23. Christopher Wills, *Exons, Introns and Talking Genes: The Science Behind the Human Genome Project* (New York: Basic Books, 1991), 10.

24. See discussion in Marga Vicedo, "The Human Genome Project," *Biology and Philosophy* 7 (1992), 255–278.

25. See Thomas J. Bouchard, Jr., David T. Lykken, Matthew McGue, Nancy Segal, and Auke Tellegen, "Sources of Human Psychological Differences: The Minnesota Study of Twins Reared Apart," *Science* 250 (12 October 1990), 223. Also Val Dusek, "Bewitching Science," *Science for the People,* November/December 1987, 19.

26. "Mapping the Genome: The Vision, the Science, the Implementation: A Roundtable Discussion," 18 February 1992, published in *Los Alamos Science* 20 (1992), 68–85.

27. Douglas Wahlsten, "Insensitivity of the Analysis of Variance to Heredity-Environment Interaction," *Behavioral and Brain Sciences* 13 (1990), 109–161.

28. Peter McGuffin and Randy Katz, "Who Believes in Estimating Heritability as an End in Itself?" 141–142, in Douglas Wahlsten, op. cit.

29. Stephen Jay Gould, "The Confusion Over Evolution," *New York Review of Books*, 19 November 1992, 48. See also Richard Lewontin, *Biology as Ideology* (New York: Harper, 1992) and Ruth Hubbard and Elijah Wald, *Exploding the Gene Myth* (Boston: Beacon Press, 1993).

30. Douglas Wahlsten, op. cit.

31. In 1992 African American groups attacked plans for a University of Maryland conference on "Genetic Factors in Crime," perceiving this as racially motivated. The controversy resulted in the withdrawal of NIH funds. See David Wheeler, "University of Maryland Conference That Critics Charge Might Foster Racism Loses NIH Support," *Chronicle of Higher Education*, 2 September 1992, A6–A8.

32. Robert Young, "Evolutionary Biology and Ideology," *Science Studies*, 1 (1971), 177–206.

33. There is a very large literature on the social negotiation of scientific knowledge, but a good starting point is Steven Shapin and Simon Schaffer, *Leviathan and the Air-Pump: Hobbes, Boyle and the Experimental Life* (Princeton, NJ: Princeton University Press, 1985). See also Bruno Latour and S. Woolgar, *Laboratory Life* (Princeton, NJ: Princeton University Press, 1986).

34. One of the clearest explications of this difficulty is Evelyn Fox Keller's essay on the pacemaker cell in the slime mold. The slime mold is a unicellular organism that can, when necessary, aggregate with other slime mold cells to form a slug and thereby crawl away to find a better place to live. In their efforts to explain this phenomenon, biologists have consis-

tently postulated a "pacemaker" cell that tells the other cells what to do. Keller argues that biological models featuring relationships of control and domination (rather than, for example, cooperation) may be especially compelling to those who expect the world to act that way. See Keller, "The Force of the Pacemaker Concept in Theories of Aggregation in Cellular Slime Mold" in her *Reflections on Gender and Science* (New Haven and London: Yale University Press, 1985), 150–157.

35. See Stephen Hilgartner for an exploration of the issue of popularization and its utility to scientists, "The Dominant View of Popularization: Conceptual Problems, Political Uses," *Social Studies of Science* 20 (1990), 519–539.

36. Quoted in Miles Orwell, *The Real Thing* (Raleigh: University of North Carolina Press, 1989), 258.

37. Marilyn Strathern, *Reproducing the Future* (New York: Routledge, 1992), 5.

38. Peggy Cooper Davis, "The Proverbial Woman," *The Record of the Association of the Bar of the City of New York* 48:1 (January/February 1993), 7–24. Davis is a legal theorist interested in how proverbial stories influence the courts. Also see Jerome Bruner et al., *A Study of Thinking* (New Brunswick, NJ: Transaction, 1986).

39. See Jeffrey Alexander, "Citizen and Enemy as Symbolic Classification," in Michele Lamont and Marcell Fournier, eds., *Cultivating Differences* (Chicago: University of Chicago Press, 1992); Bruce Kuklick, "Myth and Symbol in American Studies," *American Quarterly*, October 1992; and Jane Tompkins, *Sensational Designs: The Cultural Work of American Fiction* (New York: Oxford University Press, 1985).

40. See discussion in James B. Lemert, *Does Communication Change Public Opinion After All?* (Chicago: Nelson Hall, 1981). See also Dorothy Nelkin, *Selling Science*, 2nd ed. (New York: W. H. Freeman, 1995).

41. Allen Batteau, *The Invention of Appalachia* (Tucson: University of Arizona Press, 1990), 16.

42. Todd Gitlin, *Inside Prime Time* (New York: Pantheon, 1983).

43. Rayna Rapp, "Chromosomes and communication: The discourse of genetic counselling," *Medical Anthropology Quarterly* 2 (1988), 143–157.

44. George Lakoff and Mark Johnson, *Metaphors We Live By* (Chicago: University of Chicago Press, 1980).

45. *Signals*, Fall/Winter 1993 catalog. We thank Henrika Kurlick for bringing this to our attention.

46. See discussion in Todd Gitlin, *The Whole World is Watching* (Berkeley: University of California Press, 1980).

47. Milan Kundera, *Immortality* (New York: Harper Perennial, 1992), 114–116.

48. This term was used by Patricia Williams in discussions at the New York University Law School in 1992.

49. Zillah Eisenstein, *The Female Body and the Law* (Berkeley: University of California Press, 1988), 6.

50. Leo Buscaglia, *Love* (New York: Ballantine, 1972), 90, and *Born for Love* (New Jersey: Slack, 1992), 1.

51. "In Living Color," 23 February 1991.

52. Daniel Kevles, "Out of Eugenics: The Historical Politics of the Human Genome," in Kevles and Lee Hood, eds., *The Code of Codes: Scientific and Social Issues in the Human Genome Project* (Cambridge: Harvard University Press, 1992), 3–37.

53. On the cultural meaning of abandoned and switched babies there is some historical literature, including John Boswell, *The Kindness of Strangers: The Abandonment of Children in Western Europe from Late Antiquity to the Renaissance* (New York: Pantheon, 1988); Everett M. Ressler, *Unaccompanied Children: Care and Protection in Wars, Natural Disasters and Refugee Movements* (New York: Oxford University Press, 1988); and Barbara L. Estrin, *The Raven and the Lark: Lost*

Children in the Literature of the English Renaissance (Lewisburg, PA: Bucknell University Press, 1984).

54. Sherry Turkle, *The Second Self: Computers and the Human Spirit* (New York: Simon & Schuster, 1991), 173.

55. Statement by Ingrid Newkirk, frequently cited by animal rights activists. See James Jasper and Dorothy Nelkin, *The Animal Rights Crusade* (New York: Free Press, 1992), 46.

CHAPTER 2

1. Luther Burbank, *The Training of the Human Plant* (New York: The Century Co., 1907), 83.

2. Harry H. Cook, *Like Breeds Like* (Ontario, CA: San Aloi's Jersey Farm, 1931), 361.

3. Paul Popenoe and Roswell Hill Johnson, *Applied Eugenics* (New York: Macmillan, 1920), 29.

4. The early eugenics narratives did not refer specifically to "genes," since the concept of the gene was not yet well developed. For scientists, "factors" became "genes" in roughly the period from 1912 to 1917, but the popular eugenics literature does not commonly use the term "gene." For two studies of the development of the concept of the gene from different perspectives, see Elof Axel Carlson, *The Gene: A Critical History* (Ames: Iowa State University Press, 1989) and Lindley Darden, *Theory Change in Science: Strategies from Mendelian Genetics* (New York and Oxford: Oxford University Press, 1991).

5. The common claim that eugenics was never a popular movement—as Haller so stated in 1963, and as Philip Pauly has proclaimed anew in a 1993 review essay—overlooks the large, quirky, and tantalizing collection of popular texts we have examined. See Mark H. Haller, *Eugenics: Hereditarian Ideas in American Thought* (New Brunswick, NJ: Rutgers University Press, 1963), 177–182. Also, Philip J. Pauly, "Essay Review: The Eugenics Industry—Growth or Restructuring?" *Journal of the History of Biology* 26:1 (Spring 1993), 131–145.

6. This figure is drawn from a study of the *American Book Publishing Record, Cumulative Index 1876–1949*, based on titles, keywords, publishers, and short descriptions. There were thousands of other books published in this period that touched on eugenics—including much of the literature on family life, childrearing, and the "race question"—but we have tried to identify a particular sort of publication: intended to persuade, focused specifically on the question of eugenics, and written by a nonscientist. It has of course not been possible or practical to examine every text.

7. S. J. Holmes, *A Bibliography of Eugenics*, University of California Publications in Zoology, 25:2 (January 1924), 1–514.

8. Despite the growing historical literature on eugenics, few accounts have explored these nonscientific materials. Most historical studies of the American eugenics movement have focused on the development of eugenic ideas in the scientific literature and on the translation of these ideas into social policies such as involuntary sterilization. A notable exception is Martin Pernick's study of depictions of infant mortality in popular films and medical literature *The Black Stork: Eugenics and the Death of "Defective" Babies in American Medicine and Motion Pictures Since 1915* (New York: Oxford University Press, 1995). Pernick discusses the separate but related topic of sex education films in a short essay, "Sex Education Films, U.S. Government, 1920s," *Isis* 84:4 (1993), 766–768. Historians have been attentive to important textbooks by noted geneticists—such as William Castle's *Genetics and Eugenics*, widely used in college courses—and to the words and actions of administrators and scientists at the Eugenics Record Office at Cold Spring Harbor, New York. See Castle, *Genetics and Eugenics: A Textbook for Students of Biology and a Reference Book for Animal and Plant Breeders* (Cambridge: Harvard University Press, 1920). Historical accounts include Mark H. Haller, *Eugenics: Hereditarian Ideas in American Thought* (New Brunswick, NJ: Rutgers University Press, 1963); Daniel J. Kevles, *In the Name of Eugenics: Genetics and the Uses of Human Heredity* (New York: Knopf, 1985); Kenneth M. Ludmerer, *Genetics and American Society*

(Baltimore: Johns Hopkins University Press, 1972); Donald K. Pickens, *Eugenics and the Progressives* (Nashville, TN: Vanderbilt University Press, 1968); and Nicole Hahn Rafter, *White Trash: The Eugenic Family Studies, 1877–1919* (Boston: Northeastern University Press, 1988).

9. Folger, privately printed by the author, New York, 1919; Max Reichler (New York: Block Publishing Co., 1916); Flick (Philadelphia, J. J. McVey, 1913); Nearing (New York: Huebsch, 1912); Jordan (Baltimore: Franklin Printing Co., 1912); and Stokes (New York: C. J. O'Brien, 1917).

10. On Luther Burbank and his reputation, see Peter Dryden, *A Gardener Touched with Genius: The Life of Luther Burbank* (Berkeley: University of California Press, 1985). On Bell, see Robert V. Bruce, *Bell: Alexander Graham Bell and the Conquest of Solitude* (Boston: Little, Brown, 1973). Bell's own writings on eugenics included his 1914 *How to Improve the Race* (Washington: reprinted from the *Journal of Heredity* 5:1). See also David Starr Jordan, *The Heredity of Richard Roe: A Discussion of the Principles of Eugenics* (Boston: American Unitarian Association, 1911); J. H. Kellogg, "Needed: A New Human Race," *Proceedings of the First National Conference on Race Betterment* (1914), 431–450.

11. We would not want to argue that college course offerings can be read as a direct reflection of student interests. Yet the survey at least indicates that eugenics was widely taught. See Report of the Committee on Formal Education (C. C. Little, C. H. Danforth, H. D. Fish, H. R. Hunt, Ann Haven Morgan, E. L. Thorndike, P. W. Whiting), American Eugenics Society, Inc. (AES), 1928, in Papers of the American Eugenics Society held at the American Philosophical Society Library, Philadelphia, PA.

12. If eugenics has appeared to historians as an unsuccessful campaign by leading biologists and a few devoted propagandists, this may reflect their own priorities (interest in the works of scientists, rather than in texts produced by the "uninformed"). Kevles's account, for example, which is now the standard history of American and British eugenics, deals

with the popular meaning of eugenics in a brief (12–page) chapter that covers both American and British popularization and focuses primarily on the role of leading scientists in promoting eugenics (Kevles, 57–69). Garland Allen, who has explored the shifting meaning of eugenics and population control in the work of biologist Raymond Pearl, relegates Pearl's popular works to a footnote. Rafter's study of the accounts of degenerate families provides significant insight into these important texts, but no discussion of the larger popular world of eugenics in the United States. Kimmelman's much-cited dissertation captures one piece of the story (the meaning of eugenics in breeders' literature) but does not connect this agricultural literature to other popular materials. Our own brief foray into such uncharted waters barely begins to redress the historical gap. We do, however, want to suggest that it hints at possibilities. See Kevles, *In the Name of Eugenics: Genetics and the Uses of Human Heredity* (Berkeley and Los Angeles: University of California Press, 1985); Garland Allen, "Old Wine in New Bottles: From Eugenics to Population Control in the Work of Raymond Pearl," in Keith R. Benson, Jane Maienschein, and Ronald Rainger, eds., *The Expansion of American Biology* (New Brunswick, NJ, and London: Rutgers University Press, 1991), 231–261; Nicole Hahn Rafter, ed., *White Trash: The Eugenic Family Studies* (Boston: Northeastern University Press, 1988); and Barbara Kimmelman, "A Progressive Era Discipline: Genetics at American Agricultural Colleges and Experiment Stations" (Ph.D. diss., University of Pennsylvania, 1987).

13. G. Frank Lydston, *The Blood of the Fathers: A Play in Four Acts Dealing with the Heredity and Crime Problem* (Chicago: Riverton Press, 1912). Lydston also published a popular book on criminality, *The Diseases of Society* (Philadelphia: Lippincott, 1912), and several eugenics essays including "Some of the difficulties in the application of eugenics to the human race," *Virginia Medicine*, 14 (1909–10), 15–19. On Lombroso, the most accessible study is Stephen Jay Gould, *The Mismeasure of Man* (New York: W. W. Norton, 1981).

14. All taken from a Riverton Press plot summary, intended for reviewers, that was tucked inside Charles E. Rosenberg's personal copy of the book.

15. Lydston, 1912, 241 of text.

16. La Reine Helen Baker, *Race Improvement or Eugenics: A Little Book on a Great Subject* (New York: Dodd, Mead and Co., 1912).

17. T. W. Shannon, *Eugenics* (Marietta, OH: S. A. Mullikin Publishing Co., 1917; Garden City, NY: Doubleday and Co., 1970), 232.

18. William S. Sadler, *Race Decadence: An Examination of the Causes of Racial Degeneracy in the United States* (Chicago: A. C. McClurg and Co., 1922), 221, 246.

19. See Kevles, *In the Name of Eugenics*, 61–2. Also photo scrapbook, American Eugenics Society, Archives of the American Philosophical Society, Philadelphia.

20. Indiana University physician Thurman B. Rice said in 1929 that the "education of thinking people in the principles of race culture and hygiene is the first and most important step" in the development of a eugenic state. See *Racial Hygiene: A Practical Discussion of Eugenics and Race Culture* (New York: Macmillan, 1929), x. It is perhaps worth noting that Rice felt that one primary impediment to eugenics in the 1920s was the difficulty of acquiring complete genealogical records on each individual—for how could local authorities know how to handle a newcomer without access to complete family records? How could they decide who should be permitted to have children and who should be sterilized, if they did not know the family and its history? See *Racial Hygiene*, 313–328.

21. This according to Edwin Slossen, editor of the first science newspaper syndicate, who wrote in 1929 that the press should promote eugenics because the masses must understand how important it was. Edwin Slossen, "Democracy and Knowledge," in B. Brownell, ed., *Preface to the Universe* (New York: Van Nostrand, 1929), 108.

22. MIT biologist Frederick Adams Woods explained that in all societies the most important resource was the biological quality of the leading families; see *Mental and Moral Heredity in Royalty* (New York: Henry Holt and Co., 1906) and *The Influence of Monarchs: Steps in a New Science of History* (New York: Macmillan, 1913).

23. While several historical studies of eugenics have noted the special role that women played in the movement, both as promoters and data gatherers and as the subjects of social control, few have explored the special meaning of sex from Rafter's perspective. See Nicole H. Rafter, "Claims-Making and Socio-Cultural Context in the First U.S. Eugenics Campaign," *Social Problems* 39:1 (February, 1992), 17–33.

24. We are indebted to Hallie Levin for calling this legal citation to our attention. See Robert J. Cynkar, "Buck v Bell: Felt Necessities v Fundamental Values?" *Columbia Law Review* 81 (Nov. 1981), 1418–1461. See also Charles A. Boston, "A Protest Against Laws Authorizing the Sterilization of Criminals and Imbeciles," *Journal of the American Institute of Criminal Law and Criminology* (1913), 326–358.

25. The phrase is taken from one of the more virulently racist popular eugenics texts of the period, but it appeared in similar form in many different texts and was not original with this author. Citation is to C. S. Babbitt, *The Remedy for the Decadence of the Latin Race* (El Paso, TX: 1909), 43. Babbitt called America a "vast cesspool and dumping ground for the most degraded classes of the whole earth . . . in short, the whole world is being cleansed and America is receiving the offal."

26. Rafter, 1988, 1–31.

27. Dugdale, *The Jukes: A Study in Crime, Pauperism and Heredity* (New York: G. P. Putnam, 1910), 69–70, cited in Mark Haller, *Eugenics*, 22–23.

28. Henry Goddard, *The Kallikak Family: A Study in the Heredity of Feeblemindedness* (Cold Spring Harbor, NY: Eugenics Record Office, 1912).

29. Several historians (Rafter, 1988 particularly) have explored the questionable data-collection methods of the eugenics field workers who compiled these family histories. Field workers, for example, accepted anecdotes about deceased or absent family members as proof of pathologies. They also openly acknowledged that any individuals in the families who lived "healthy and industrious lives" were not included in the studies.

30. Preface by Charles B. Davenport to Anna Wendt Finlayson, *The Dack Family: A Study in the Hereditary Lack of Emotional Control* (Cold Spring Harbor, NY: Eugenics Record Office, Bulletin 15, 1916), v–vi.

31. This was a theme in many of the family stories, in which the environment was interpreted as a consequence of the biological flaws of the families. See Popenoe and Johnson, *Applied Eugenics,* 168.

32. "We have two clear matings of industry and laziness where the lazy parent (mother) had two lazy parents [and] of 10 offspring 9 were lazy and only 1 industrious like the father." Arthur H. Estabrook and Charles Davenport, *The Nam Family: A Study in Cacogenics* (Cold Spring Harbor, NY: Eugenics Record Office, Memoir No. 2, August 1912), 66–67.

33. Ibid., 74.

34. For a full report on the 1914 conference, see Race Betterment Foundation, *Proceedings of the First National Conference on Race Betterment,* January 1914, 597.

35. The society also sponsored sermon contests, public lecture series, and popular exhibits. Details of the public education program of the American Eugenics Society can be found in the AES papers of the American Philosophical Society, Philadelphia. See also List of Eugenics Speakers and Sermon Contest files, Papers of the AES, APS, Philadelphia, PA.

36. Race Betterment Foundation, *Proceedings of the First National Conference on Race Betterment,* January 1914, 621.

37. There were "enough children of each age" who were "worthy of the gold medals for their teeth, but lack of other

qualities caused them to fail to appear in the competition."
Ibid., 608.

38. "There is nothing which has occurred during this conference which has attracted more attention and in which more
interest has been taken, especially by fathers and mothers,
brothers and sisters, than in this baby contest," said a Battle
Creek, Michigan, jurist whose own son was entered in the
contest. Ibid., 624, 599.

39. These competitions began in 1920 at the Kansas Free
Fair in Topeka and quickly became popular; by 1930, fitter
family contests were featured at more than forty state fairs
each year. See Kevles, *In the Name of Eugenics*, 61–62. One
chart displayed at such fairs stated that "very few *normal*
persons ever go to jail." Photo scrapbook, American Eugenics Society, Archives of the American Philosophical Society,
Philadelphia.

40. "Fitter Families Examination," Eugenics Society of the
United States of America. Form with instructions to examiners in papers of the American Eugenics Society, Archives of
the American Philosophical Society, Philadelphia.

41. Viviana A. Zelizer, *Pricing the Priceless Child: The Changing Social Value of Children* (New York: Basic Books, 1985).
She tracks a shift in the meaning of children (roughly
1925–1935) as "children's accidental death emerged as a
'serious, fundamental national problem,'" 37.

42. For a provocative analysis of the historical meaning of
the Scopes trial, see Susan Harding, "Representing Funda-
mentalism: The Problem of the Repugnant Cultural Other,"
Social Research 58:2 (Summer 1991), 373–393. See also Ray
Ginger, *Six Days or Forever* (Boston: Beacon Press, 1958) and
Jerry R. Tompkins, ed., *D-Days at Dayton: Reflections on the
Scopes Trial* (Baton Rouge: Louisiana State University Press,
1965).

43. Edward Bagby Pollard, *The Rights of the Unborn Race*
(Philadelphia and Boston: Social Service Series, American
Baptist Publication Society, 1914). A similar publication by
the Canadian Purity Education Association (CPEA) in 1911

featured a baby, in the care of a stork, who refused to be delivered to parents who were not properly trained in eugenics. See CPEA, *Baby's Rights* (promotional pamphlet, 1911).

44. American Eugenics Society, *A Eugenics Catechism* (1926) in the papers of the American Eugenics Society, Archives of the American Philosophical Society, Philadelphia.

45. See American Eugenics Society, *A List of Eugenics Lecturers* (1927) in the Papers of the American Eugenics Society, Archives of the American Philosophical Society, Philadelphia.

46. The question of eugenics and its appropriation is explored in most of the historical studies cited above; see note 21 to Ch. 2, but particularly Mark H. Haller, *Eugenics: Hereditarian Attitudes in American Thought* (New Brunswick, NJ: Rutgers University Press, 1963).

47. We are indebted to Diane Paul for pointing out the combination of messages in this film when she viewed it with us at a course in the history of human genetics held at Woods Hole, Massachusetts, in August 1993.

48. Robert Proctor, *Racial Hygiene: Medicine Under the Nazis* (Cambridge: Harvard University Press, 1988).

49. For what is still one of the best accounts of this process, see Haller, *Eugenics*, 76–94 and 177–189.

50. See Chapter 7. Also Lynn Preston Copeland, "How Science Is Charting the Way to Happiness for Your Child," *Parents*, May 1950, 38–40, 119.

51. Lucien Malson, *Wolf Children and the Problem of Human Nature* (New York and London: Monthly Review Press, 1972), 9.

52. Pendleton Tompkins, "A New Journal," *Fertility and Sterility* 1 (1 January 1950), 1–2.

53. Alan F. Guttmacher, "Artificial Insemination," *Fertility and Sterility* 5:1 (1954), 4.

54. Those sitting around the conference table included geneticist and physician James V. Neel (one of the most important figures in the rise of scientific human genetics in the United States), the physician-turned-geneticist Victor McKusick; human geneticists Arno Motulsky and William J. Schull; fly geneticists Curt Stern, James F. Crow, and Bentley Glass; and bacterial geneticist Joshua Lederberg. See William J. Schull, *Genetic Selection in Man: Third Macy Conference on Genetics* (Ann Arbor: University of Michigan Press, 1963).

55. Ibid.

56. Victor McKusick, *Human Genetics* (Englewood Cliffs, NJ: Prentice Hall, 1964), 4.

57. Osborne, Frederick, "History of the American Eugenics Society," *Social Biology* 20 (1974), 115–126.

58. Rafter, *White Trash*, 27–28.

59. See discussion of this literature in Chapter 9.

60. Popenoe and Johnson, *Applied Eugenics*, 29.

CHAPTER 3

1. Bryan Appleyard, in his *Understanding the Present: Science and the Soul of Modern Man* (New York: Doubleday, 1992), describes this tendency as an effort to establish a "church" of science and to give scientific findings moral and cultural force. Ch. 1 and *passim*.

2. Stephen Hawking, *A Brief History of Time* (New York: Bantam, 1988).

3. Steven Weinberg, *Dreams of a Final Theory* (New York: Pantheon, 1993).

4. Cited in Bryan Appleyard, "In Science We Trust," *New York Times* op-ed, 7 April 1993.

5. Leon Lederman, *The God Particle* (New York: Houghton Mifflin, 1992).

6. The synthesis is the focus of considerable historical interest. See, for example, V. B. Smocovitis, "Unifying Biology: The Evolutionary Synthesis and Evolutionary Biology," *Journal of the History of Biology* 25:1 (1992) and Jonathan Harwood, "Metaphysical Foundations of the Evolutionary Synthesis: A Historiographical Note," *Journal of the History of Biology* 27:1 (1994).

7. See Lily E. Kay, *The Molecular Vision of Life: Caltech, the Rockefeller Foundation and the Rise of the New Biology* (New York and Oxford: Oxford University Press, 1993), esp. 143–163.

8. E. O. Wilson, *Sociobiology: The New Synthesis* (Cambridge: Harvard University Press, 1975). See also R. C. Lewontin, S. Rose, and L. Kamin, *Not in Our Genes* (New York: Pantheon, 1984).

9. L. Jaroff, "The Gene Hunt," *Time*, 20 March 1989, 62–71.

10. Gregory Stephen Henderson, "Is DNA God?" *The Pharos*, Journal of the Alpha Omega Alpha Honor Medical Society, Winter 1988, 2–6.

11. These quotations are taken from A. E. Crawley's 1909 study (dedicated to eugenicist Francis Galton) *The Idea of the Soul* (London: Adam and Charles Black), 209, 211, 239.

12. *New Catholic Encyclopedia*, s. v. "soul, human," 1967. Also Mircea Eliade, *The Sacred and the Profane* (New York: Harcourt Brace, 1957).

13. Swinburne, *The Evolution of the Soul* (Oxford: Clarendon Press, 1986). See also his discussion of soul and identity and their relation to both physical appearance and fingerprinting, which establishes identity by "indirect evidence," 161–163.

14. Eliade, *Sacred*, 11.

15. Carolyn Bynum, "Material Continuity, Personal Survival and the Resurrection of the Body: A Scholastic Discussion in Its Medieval and Modern Contexts," in her *Fragmentation and Redemption: Essays on Gender and the Human Body in Medieval Religion* (New York: Zone, 1991), 239–297.

16. Gary Bergel, "When You Were Formed in Secret," (Reston, VA: National Right to Life pamphlet, 1988), 2.

17. Sherry Turkle, "Artificial Intelligence," *Daedalus*, Winter 1988, 117.

18. Howard Rheingold, *Virtual Reality* (New York: Simon and Schuster, 1991).

19. James Jasper and Dorothy Nelkin, *The Animal Rights Crusade* (New York: Free Press, 1992).

20. Melvin Konner, *Why the Reckless Survive and Other Secrets of Human Nature* (New York: Penguin Books, 1990) and Richard Dawkins, *The Selfish Gene* (New York: Oxford University Press, 1976).

21. Andrew Ross, *Strange Weather: Culture, Science and Technology in the Age of Limits* (London: Verso, 1991), 152.

22. Donna Haraway, *Simians, Cyborgs and Women: The Reinvention of Nature* (New York: Routledge, 1991).

23. J. Madeleine Nash, "Copying what comes naturally: Scientists are creating revolutionary new materials by imitating the weave and structure of biological designs," *Time*, 8 March 1993, 58–59.

24. Bryan Appleyard, *Understanding the Present*, cover.

25. Abby Lippman, "SimDisaster," *Hastings Center Report*, March-April 1993, 3–4.

26. There is a massive literature in film studies journals analyzing *Blade Runner*, but our perspective on the film is quite different. For a sampling, see the extensive list of citations in W. M. Kolb, "*Blade Runner*: An Annotated Bibliography," *Literature/Film Quarterly* 18:1 (1990), 19–64. See also Yves Chevrier, "*Blade Runner*, or the Sociology of Anticipation," *Science Fiction Studies* 11 (1984), 50–60; David Dresser, "*Blade Runner*: Science Fiction and Transcendence," *Literature/Film Quarterly* 14:2 (1985), 89–100; and Per Schelde, *Androids, Humanoids and Other Science Fiction Monsters: Science and Soul in Science Fiction Films* (New York: New

York University Press, 1993). We are indebted to Douglas Keen for calling this literature to our attention.

27. Mark Evanier et al., *DNAgents*, Eclipse Enterprises series; "X-Cutioner's Song—Part I" in *Uncanny X-Men* 1:294 (November 1992), 19. We are indebted to Daniel Gilman for sharing his comic books with us.

28. Melvin Konner, *Why the Reckless Survive*, 223–224.

29. Eric Lander, "DNA Fingerprinting: Science, Law and the Ultimate Identifier," in Daniel Kevles and Leroy Hood, *The Code of Codes* (Cambridge: Harvard University Press, 1992), 193.

30. Anastasia Toufexis, "Convicted by Their Genes," *Time*, 31 October 1988, 74; and "DNA Prints: A Foolproof Crime Test," *Time*, 26 January 1987, 66.

31. For discussion of the uncertainties, see Eric Lander, "DNA Fingerprinting," 191–210.

32. Joan E. Rigdon, "DNA Expert Peers at Crooks, Soldiers, Little Bighorn Bones," *Wall Street Journal*, 6 January 1994.

33. The exhibit was "Genus Memorialis: Imitations of Immortality" at Horodner Romley Gallery in Manhattan, February 1993. In December 1994, artist Suzanne Anker curated an exhibit called "Gene Culture: Molecular Metaphor in Visual Art" at Fordham University, New York.

34. Camille Paglia, *Sex, Art and American Culture* (New York: Vintage, 1992), 103–104.

35. Rick Weiss, "Techy to Trendy, new products hum DNA's tune," *New York Times*, 8 September 1992.

36. N. A., "Kary Mullis," *Omni*, April 1992, 69–92.

37. Rick Weiss, "Techy to Trendy."

38. *New Catholic Encyclopedia*, "relics," 1967.

39. *New Catholic Encyclopedia*, 1967. In the ninth century, a corporation was formed to discover, sell, and transport relics throughout Europe. False relics proliferated, particularly in

the wake of the Crusades in the thirteenth century. The commercial practices that built up around relics became a focus of Protestant satire: John Calvin's (1509–1564) treatise on relics said the bits of wood from the "True Cross" were so numerous and heavy that not even three hundred men could have carried such a cross.

40. See Peter Brown, *The Cult of the Saints: Its Rise and Function in Latin Christianity* (Chicago: University of Chicago Press, 1981), 89. Hinduism and Judaism have both rejected relics, possibly because in both traditions the bodies of the dead are interpreted as impure. But Buddhism and Roman Catholicism have accepted and even endorsed relics.

41. Warren E. Leary, "Scientists Seek Lincoln DNA to Clone for a Medical Study," *New York Times*, 10 February 1991.

42. Leslie Roberts, "Genome Diversity Project: Anthropologists Climb (Gingerly) on Board," *Science* 258 (20 November 1992), 1300–1301. See also the Senate Committee on Government Affairs, U.S. Congress, "Hearings on the Human Genome Diversity Project," 26 April 1993.

43. Dawkins, *The Selfish Gene*, 24, 36.

44. Paul Kroll, "The Gene Healers: Curing Inherited Diseases," *The Plain Truth: A Magazine of Understanding* 55:8 (September 1990), 3–8.

45. For a history of this controversy, see Sheldon Krimsky, *Genetic Alchemy: The Social History of the Recombinant DNA Controversy* (Cambridge: MIT Press, 1982).

46. Stephen R. Donaldson, "Animal Lover," reprinted in Donaldson, *Daughter of Regals and Other Tales* (New York: Ballantine Books, 1984), 247, 260.

47. Robin Cook, *Mutation* (New York: Putnam, 1989).

48. Michael Stewart, *Prodigy* (New York: HarperCollins, 1991).

49. Philip Elmer Dewitt, "The Perils of Treading on Heredity," *Time*, 20 March 1989, 70. Illustration by Joe Lertola.

50. Cartoon by Stuart Goldenberg, *New York Times,* 16 September 1990.

51. Andrew Kimbrell, *The Human Body Shop* (San Francisco: Harper, 1993). Also see Daniel Callahan, "They Dream of Genes," *New York Times Book Review,* 12 September 1993.

CHAPTER 4

1. Marilyn Strathern and Sarah Franklin, "Unanticipated Contexts: The Representation of Kinship in the Context of New Reproductive Technologies," *Anthropology in Action,* Spring 1992, 4.

2. Marilyn Strathern, *Reproducing the Future* (New York: Routledge, 1992).

3. David M. Schneider, *American Kinship: A Cultural Account* (Englewood, NJ: Prentice Hall, 1968).

4. The proliferation of such references reflected the "family values" rhetoric in the political arena.

5. *In re Allison D.,* 572 N.E.2d 27, 28 (N.Y. 1991); the court denied parental rights to the former partner of a lesbian mother despite a continuous six-year relationship with the child because she was a "biological stranger" to the child.

6. See Sarah Franklin, "Deconstructing Desperateness," in Maureen McNeil, Ian Varnoe, and Steven Yearly, eds., *The New Reproductive Technologies* (New York: St. Martin's Press, 1990), 207.

7. *Look Who's Talking* (United States: Tri-Star, 1989).

8. For dozens of such stories, see Susan Faludi, *Backlash: The Undeclared War Against the American Woman* (New York: Crown Publishers, Inc., 1992).

9. Anne Taylor Fleming, *Motherhood Deferred: A Woman's Journey* (New York: G. P. Putnam, 1994).

10. Carroll Smith-Rosenberg and Charles Rosenberg, "The Female Animal: Medical and Biological Views of Woman and

Her Role in Nineteenth-Century America" in Judith Walzer Leavitt, ed., *Women and Health in America* (Madison: University of Wisconsin Press, 1984), 12–27.

11. See Willy de Craemer, "A Cross-Cultural Perspective on Personhood," *Milbank Memorial Fund Quarterly* 61:1 (1983), 23.

12. Judith Lasker and Susan Borg, *In Search of Parenthood: Coping with Infertility and High Tech Contraception* (Boston: Beacon Press, 1987), 12, 20.

13. Quoted in Sarah Franklin, "Deconstructing Desperateness," 207.

14. Ann Hood, "Hatching Hope," *Redbook*, April 1992, 58.

15. Letter to the editor, *Resolve*, April 1989, 2.

16. Susan Faludi, *Backlash*, 105–106.

17. David Riley, "The Quest for a Child," *Parents*, April 1980, 120–129.

18. IVF America Inc., a chain of fertility clinics, announced its planned initial public offering in spring 1992 at the asking price of $8 to $10 a share. "Can a Baby Making Venture Deliver," *New York Times*, 1 June 1992.

19. Ellen Hopkins, "Tales from the Baby Factory," *New York Times Magazine*, 15 March 1992, 80.

20. Ibid., 40–41.

21. Jean Marzollo, "Confessions of a (Sort of) Grown-up," *Parents*, July 1981, 47.

22. Susan Edmiston, "Whose Child Is This?" *Glamour*, November 1991, 236.

23. Lynn Preston Copeland, "How Science Is Charting the Way to Happiness for Your Child," *Parents*, May 1950, 38–40, 119.

24. The importance of fetal environment in such debates reflects the intricacies of worker's compensation laws in many states. Workers, if injured, are compensated according

to a fixed schedule without regard to fault, but they cannot directly sue their employers. The fetus, as a third party, can sue. Thus employers are concerned with protecting the fetus in order to prevent future lawsuits.

25. "As the World Turns," 28 May 1991–July 1992.

26. John Poppy, "Healthy Roots," *Esquire,* November 1990, 73.

27. Betty Jean Lifton, *Twice Born: The Adoption Experience* (New York: Harper, 1979) and *Journey of the Adopted Self: A Quest for Wholeness* (New York: Basic Books, 1994).

28. For the increase in public confessions, see Wendy Kaminer, *I'm Dysfunctional, You're Dysfunctional* (New York: Addison Wesley, 1992).

29. John McCormack and Pat Wingert, "Whose Child Is It Anyway?" *Newsweek,* Special Issue, summer 1991, 58.

30. Katherine Bishop, "For the Adopted the Issue of Roots Grows Stronger," *New York Times,* 7 October 1990.

31. Cited in "Tune in to Adoption Myths," *Psychology Today,* November 1988, 12.

32. Mona Simpson, *The Lost Father* (New York: Knopf, 1991).

33. Leah D. Frank, "Woman Searches for Birth Mother," *New York Times,* 5 May 1991.

34. "The Simpsons," 21 February 1990.

35. Richard Weizel, "First Person: A Voice from the Past," *Parenting,* October 1991, 90.

36. "In Search of the Fathers," WGBH (Boston), October 1992.

37. Walter Goodman, "Giving Up a Child and Being Given Up," *New York Times,* 28 November 1990.

38. Judith Lasker and Susan Borg, *In Search of Parenthood* (Boston: Beacon Press, 1987), 158.

39. Kenneth Kaye, "Turning Two Identities into One," *Psychology Today*, November 1980, 46–50.

40. Patrick McGilligan, review of *Walt Disney*, by Marc Eliot, *New York Times Book Review*, 18 July 1993.

41. Lasker and Borg, *In Search of Parenthood*, 14.

42. Kenneth Kaye, "Turning Two Identities into One," *Psychology Today*, November 1980, 46–50.

43. Tamar Lewin, "Adopted Youths Are Normal in Self-Esteem, Study Finds," *New York Times*, 23 June 1994. Report of a study of 715 adoptive families by the Search Institute in Minneapolis.

44. Sherry Bunin, "Up with Adoption," *Parents*, January 1990, 78–80.

45. These and similar statements appear throughout *Resolve*, a newsletter for infertile people (P.O. Box 175, Gracie Station, New York).

46. Susan Edmiston, "Whose Child Is This?" *Glamour*, November 1991, 235.

47. Elizabeth Bartholet, *Family Bonds* (New York: Houghton Mifflin, 1993), 59.

48. Ibid., 167.

49. Lucinda Franks describes these groups in "The War for Baby Clausen," *The New Yorker*, 22 March 1993, 56–73.

50. *A Guide to Selected National and Genetic Voluntary Organizations* (Washington D.C.: National Center for Education in Maternal and Child Health, January 1989).

51. Florence Anna Fisher, letter to the editor of the *New York Times*, 9 May 1993.

52. Quoted in John Riley, "Adoption Rules Changing," *Newsday*, 17 May 1993.

53. Cover of *Time*, 15 August 1994, linked to a story by Robert Wright, "Our Cheating Hearts," 44–52.

54. George Lipsitz, *Time Passages: Collective Memory and Popular Culture* (Minneapolis: University of Minnesota Press, 1990) and Ella Taylor, *Prime-Time Families* (Berkeley: University of California Press, 1989).

55. Taylor, *Prime-Time Families*, 65.

56. Ibid., 158.

57. "Pop the Question," ibid.

58. "The Guiding Light," May 1991–July 1992.

59. "All My Children," May 1991–July 1992.

60. "Switched at Birth," 28 April 1991.

61. Quoted in Larry Eohter, "Natural vs Adoptive," *New York Times*, 30 July 1990.

62. Isabel Wilkerson, "Custody Battle: Is Conception Parenthood?" *New York Times*, 27 December 1992, 20.

63. Editorial, *New York Times*, 23 August 1993. The Iowa judge felt obliged to recognize the biological mother's claim: any other decision, he said, would engage the court in social engineering. Responding to an emergency appeal, Supreme Court Justice Paul Stevens concurred. But Justices Blackmun and O'Connor dissented, worrying that the laws ignored the personal vulnerability of the child: "Court Says Girl Must Return to Biological Parents," *New York Times*, 31 July 1993. For a summary of the case, see Lucinda Franks, "The War for Baby Clausen," 56–73.

64. Lynda Richardson, "A Call Unmade and Siblings Split," *New York Times*, 7 August 1992.

65. Family Tree Maker, Banner Blue, PO Box 7865, Fremont, CA 94537.

66. Len Hilts, *How to Find Your Own Roots* (Illinois: Great Lakes Living Press, 1977).

67. Richard Phalon, "Great-great-grandfather Was a What?" *Forbes*, April 1991, 122.

68. James Gorman, "The DNA of the DAR," *Discover,* May 1985, 62.

69. Elizabeth Moore, "Growing a Family Tree," *Parenting,* November 1991, pull-out section.

70. *Time,* 15 August 1994, cover.

71. Carol Levine, "AIDS and Changing Concepts of Family," in Dorothy Nelkin, David Willis, and Scott Parris, eds., *A Disease of Society* (New York: Cambridge University Press, 1991), 45–70.

72. Martha Field, *Surrogate Parenting* (Cambridge: Harvard University Press, 1988).

73. Robert Wright, *The Moral Animal: Evolutionary Psychology and Everyday Life* (New York: Pantheon, 1994) and Robert Wright, "Our Cheating Hearts," 44–52. Also see David Buss, *Evolution of Desire* (New York: Basic Books, 1994); William Allman, *Psycho Darwinism* (New York: Harper-Collins, 1994); and Helen E. Fisher, *Anatomy of Love* (New York: Fawcett Columbine, 1992).

CHAPTER 5

1. Elaine Dundy, *Elvis and Gladys* (New York: St. Martin's Press, 1985), 26. See discussion in Greil Marcus, *Dead Elvis: A Chronicle of a Cultural Obsession* (New York: Doubleday, 1991).

2. Dundy, *Elvis and Gladys,* 74.

3. Albert Goldman, *Elvis* (New York: McGraw-Hill, 1981), 57. See discussion in Greil Marcus, *Dead Elvis.*

4. See, for example, Alan Wexler, "Everything into the Pool," *Newsday,* 13 August 1993.

5. See special issue of *Science* 264 (17 June 1994), 1685–1739.

6. Lisa Lynch, "Twins Personality Studies Made Big Splash Before Journal Approval," *Science Writer,* Winter 1990, 15–20.

7. *U.S. News and World Report,* 13 April 1987, 163.

8. These examples are from Lynch, "Twins Personality Studies."

9. Madeline Drexler, "Geneticists Have Ascendancy," *Boston Globe,* 3 December 1989.

10. Thomas J. Bouchard et al., "Sources of Human Psychological Differences: The Minnesota Studies of Twins Reared Apart," *Science* 250 (12 October 1990), 223.

11. Lynch, "Twins Personality Studies."

12. Newspaper searches were facilitated by Nexus, made accessible to the authors through the NYU School of Law.

13. See, for example, the anthropological literature on witchcraft, e.g. Mary Douglas, *Purity and Danger* (London: Routledge, 1966).

14. Stanley Milgram, "Behavioral Studies of Obedience," *Journal of Abnormal and Social Psychology* 67 (1963), 371–378; and B. F. Skinner, *Beyond Freedom and Dignity* (New York: Knopf, 1971).

15. P. A. Jacobs, M. Bruton, M. M. Melville, R. P. Brittan, and W. F. McClement, "Aggressive Behavior, Mental Subnormality and the XYY Male," *Nature* 208 (1965), 1351–1352.

16. Cited in Jeremy Green, "Media Sensationalism and Science: The Case of the Criminal Chromosome," in Terry Shinn and Richard Whitley, eds., *Expository Science,* Sociology of the Sciences Yearbook IX (1985), 139–161. Green provides many examples of the generalizations conveyed in both scientific and popular accounts.

17. Ibid., 144.

18. *New York Times,* 23 April 1986.

19. Richard Hutton and George Page, "The Mind/The Brain Classroom Series," *PBS* video, 1992.

20. "The Donahue Show," 25 February 1993. The program is described in John Horgan, "Eugenics Revisited," *Scientific American*, June 1993, 123.

21. Maria Newman, "Raising Children Right Isn't Always Enough," *New York Times*, 22 December 1991.

22. James Fallows, "Born to Rob? Why Criminals Do It," *The Washington Monthly*, December 1985, 37.

23. George Nobbe, "Alcoholic Genes," *Omni*, May 1989, 37.

24. "The Oprah Winfrey Show," 24 August 1992.

25. *JFK*, Warner Bros., 1991.

26. Deborah Franklin, "What a Child Is Given," *New York Times Magazine*, 3 September 1989, 36.

27. Camille Paglia, *Sex, Art, and American Culture* (New York: Vintage Books, 1992), 105.

28. *The Atlanta Journal and Constitution*, 17 April 1991.

29. Ibid., 6 February 1991.

30. Nick Downes, cartoon in *Science* 256 (24 April 1992), 547.

31. James O. Jackson, "The New Germany Flexes Its Muscles," *Time*, 13 April 1992, 34.

32. Christopher Lehman-Haupt, "Studying Soccer Violence by the Civilized British," *New York Times*, 25 June 1992.

33. Maria Newman, "Raising Children Right."

34. *Tainted Blood*, USA Channel, 3 March 1993.

35. Deann Glamser, "Killer's Brain Causes Clash," *USA Today*, 6 January 1993.

36. Fox Butterfield, "Studies Find a Family Link to Criminality," *New York Times*, 31 January 1992.

37. Daniel Koshland, "Elephants, Monstrosities, and the Law," *Science* 255:5046, 14 February 1992, 777.

38. Sarnoff Mednick, "Crime in the Family Tree," *Psychology Today*, March 1985, 17.

39. Melvin Konner, "The Aggressors," *New York Times Magazine*, 14 August 1988, 33–34.

40. Jeremy Green, "Media Sensationalism and Science," 139–161.

41. L. L. Coleman, "Speak of Your Health," reprinted in Jeffrey Goldstein, *Seville Statement on Violence* (unpublished document), November 1990, 182. Note that Watson and Crick did not study or comment on XYY males.

42. National Academy of Sciences, National Research Council, *Understanding and Preventing Violence* (Washington, DC: National Academy Press, November 1992).

43. Fox Butterfield, "Study Cites Role of Biological and Genetic Factors in Violence," *New York Times*, 13 November 1992.

44. Anastasia Toufexis, "Seeking the Roots of Crime," *Time*, 19 April 1993, 52–53.

45. R. A. Baron and D. Byrne, *Social Psychology* (Boston: Allyn and Bacon, 1991), 202.

46. Michael Ghiglieri, "War Among the Chimps," *Discover*, November 1987, 66.

47. See, for example, Marvin Harris, *Cows, Pigs, Wars and Witches: The Riddle of Culture* (New York: Random House, 1974); David Barash, *The Whisperings Within* (New York: Harper and Row, 1979), and Konrad Lorenz, *On Aggression* (New York: Bantam, 1967).

48. Jeffrey Goldstein, *Seville Statement*.

49. The *Seville Statement* and list of signatories is included in Anne E. Hunter, ed., *Genes and Gender VI: On Peace, War and Gender* (New York: Feminist Press, 1991), 168–171.

50. Jeffrey Goldstein, *Seville Statement.*

51. Peter Conrad and Joseph Schneider, eds., *Deviance and Medicalization* (St. Louis: C. V. Mosby, 1980).

52. Sheila B. Blume, M.D., "The Disease Concept of Alcoholism," *Journal of Psychiatric Treatment and Evaluation* 5 (1983), 417–478; David Musto, *The American Disease* (New Haven, CT: Yale University Press, 1973), and Joseph Gusfield, *The Culture of Public Problems* (Chicago: University of Chicago Press, 1981).

53. Republished in 1960 as E. M. Jellinek, *The Disease Concept of Alcoholism* (Highland Park, NJ: Hillhouse Press). See also Sheila B. Blume, M.D., "The Disease Concept of Alcoholism."

54. Conrad and Schneider, *Deviance.*

55. George Nobbe, "Alcoholic Genes."

56. Shifra Diamond, "Drinking Habits May Be in the Family," *Mademoiselle*, August 1990, 136.

57. Editorial, "Just Blame Genes—of Disease," *Christian Science Monitor*, 22 May 1991.

58. Daniel Goleman, "Scientists Pinpoint Brain Irregularities in Drug Addicts," *New York Times*, 26 June 1990.

59. See, for example, Irish Hall, "Diet Pills Return as Long-Term Medication Not Just Diet Aids," *New York Times*, 14 October 1992, and Gina Kolata, "Where Fat Is a Problem Heredity Is the Answer," *New York Times*, 24 May 1990.

60. Laura Mansnerus, "Smoking: Is It a Habit or Is It Genetic?" *New York Times Magazine*, 4 October 1992.

61. "Dear Abby" column, "Smoker Has No Plans to Kick Lifelong Habit," *Delaware County Times*, 7 April 1993, 49.

62. Lee Bar, Ph.D., *Getting Control—Overcoming Your Obsessions and Compulsions* (New York: Plume, 1992), 12–13.

63. Linda Berman and Mary Ellen Sigel, *Behind the 8-Ball: A Guide for Families of Gamblers* (New York: Simon and Schuster, 1992), 43–44.

64. Richard C. Lewontin, *Biology as Ideology* (New York: Harper and Row, 1992), 51.

65. Bruce Weber, "Chess Moves Are Planned, Birthdays Happen," *New York Times*, 5 August 1992.

66. "Kids and Stress," NBC News Special, 25 April 1988.

67. David Gelman, "The Miracle of Resiliency," *Newsweek* special issue, summer 1991, 44-47.

68. Barbara Delatiner, "For Brothers, Poetry Is in Their Genes," *New York Times*, 26 May 1991.

69. Mervyn Rothstein, "Isaac Asimov, Whose Thoughts and Books Traveled the Universe, Is Dead at 72," *New York Times*, 7 April 1992.

70. Richard Stengel, "The Swiss School of Rap," *New York Times*, 18 October 1992.

71. Robert Kennedy, *Beef It: Upping the Muscle Mass* (New York: Sterling Publishing Co., 1983).

72. Maria Terrone and Sharon Johnson, "Fashion's Nature vs Nurture Debate, Or Is Good Taste in the Genes?" *New York Times*, 12 April 1992.

73. Quoted in Felicia Ley, "Pride and Some Regrets," *New York Times*, 1 January 1994.

74. Peter Costello, *James Joyce: The Years of Growth* (New York: Pantheon, 1993); Christopher Lehmann-Haupt observed the focus on genetics in his review, *New York Times*, 8 April 1993.

75. Rob Brezsny, "Real Astrology," *New York Press*, 5-11 August 1992, 54.

76. "After I began dabbling with flowers as an adult, I was astonished to realize my dad had left me a legacy . . . his talents as a gardener had trickled down to me": in an essay on gardening called "I Found It in My Other Genes" by Martha Smith in her *Beds I Have Known: Confessions of a Passionate Amateur Gardener* (New York: Atheneum, 1990), 17.

77. *Time*, 5 November 1990.

78. Advertisement, BMW Inc. of North America, 1983.

79. *New York Times Magazine* during May 1991.

80. *New Yorker*, 5 October 1992.

81. *Vogue*, October 1991.

82. *Mirabella*, January 1993. The bottle pictured in some ads for Bijan's DNA perfume has the amazing shape of a triple helix.

83. Advertisement, Calvin Klein, cited in Anne Fausto-Sterling, *Myths of Gender* (New York: Basic Books, 1985), 7.

84. Edwin Diamond, "Can You Change a Magazine's DNA?" *New York Magazine*, 20 July 1991, 27.

85. Pascale Le Draoulec, "The 139 Children of Dr. Graham," *California Magazine*, September 1991, 46.

86. "Confessions of a Sperm Donor," *Self*, November 1991, 152.

87. "The Young and the Restless," June 1991–July 1992.

88. "Santa Barbara," June 1991–July 1992.

89. "Doogie Houser, M.D.," 1992.

90. Steve Futterman, "Hot Jazz Artists," *Rolling Stone*, 13 May 1993, 72.

91. "The Arsenio Hall Show," 2 August 1992.

92. NBC "Today Show," 21 October 1992.

93. Scott Hamilton reporting on the Olympic women's ice-skating competition, "CBS Olympic Coverage," 21 February 1992.

94. William Safire, "Dollie and Johnny," *New York Times*, 7 September 1992.

95. "Genius: The Myths and Reality," *Skeptic* 2:1 (1993), 40–52.

96. Anthony Lewis, "Politics and Decency," *New York Times,* 4 April 1991.

97. Lawrence Wright, "The Man from Texarkana," *New York Times Magazine,* 28 June 1992.

98. Steven A. Holmes, "For Buchanan Aide, Genetic Conservatism," *New York Times,* 7 February 1992.

99. Alessandra Stanley, "When Ms. Right Falls for (Gasp!) Mr. Left," *New York Times,* 20 April 1992.

100. Anna Quindlen, "Wives for Wives' Sake," *New York Times,* 9 August 1992.

101. Diane Cole, "The Entrepreneurial Self," *Psychology Today,* June 1989, 60.

102. Deborah Sontag, "Would-be Owner of the Post Takes Delight in Controversy," *New York Times,* 17 March 1993.

103. The link between creativity and madness is an idea at least as old as Aristotle, who saw poets as "clearly melancholic." These links were elaborated in the nineteenth century in such works as Cesare Lombroso's *Genius and Madness,* which explored the psychopathological origins of creativity. Today the belief that the mind is shaped by the biology of the brain has revived this theme in both professional discussions and the popular media, where talent is often associated with alcoholism and genius with mental instability. See Kay Redfield Jamison, *Touched with Fire: Manic-Depressive Illness and the Artistic Temperament* (New York: Free Press, 1993).

104. Jeannie Park and Robin Micheli, "Falling Down . . . and Getting Back Up Again," *People,* 29 January 1990, 57, and Joanne Kaufman, "Drew Barrymore," *Ladies Home Journal,* March 1990, 116.

105. Bernard Weinraub, "The Name is Barrymore," *New York Times,* 7 March 1993.

CHAPTER 6

1. Alison Jagger, *Feminist Politics and Human Nature* (Sussex: Harvester Press, 1983). See also Deborah L. Rhodes, "The No-Problem Problem: Feminist Challenges and Cultural Change," *Yale Law Journal* 100:1 (1991), 1–62.

2. Jeffries, who has published little scholarly work, is well known because of his work with the New York City Board of Education on curricular reform and his public speeches in the late 1980s and early 1990s. See Eric Pooley, "Doctor J: The Rise of Afrocentric Conspiracy Theories. Leonard Jeffries and His Odd Ideas about Blacks and Whites," *New York Magazine*, 2 September 1991.

3. Simon LeVay, *The Sexual Brain* (Cambridge: MIT Press, 1993).

4. Christine Gorman, "Sizing Up the Sexes," *Time*, 20 January 1992, 42.

5. Ibid.

6. Daniel Gelman, "Just How the Sexes Differ," *Newsweek*, 18 May 1981, 2–28.

7. Elaine Morgan, *The Descent of Woman* (New York: Stein and Day, 1972).

8. E. O. Wilson, *Sociobiology* (New York: Cambridge University Press, 1975), and E. O. Wilson, "Human Decency Is Animal," *New York Times Magazine*, 12 October 1975.

9. Richard Dawkins, *The Selfish Gene* (New York: Oxford University Press, 1976), 151–178.

10. Phyllis Schlafly, *The Power of the Positive Woman (1977)*, 12–17. Cited in Rhodes, "The No-Problem Problem."

11. Richard Berke, "Religious Right Gains Influence," *New York Times*, 3 June 1994.

12. Camille Benbow and Julius Stanley, "Sex Differences in Mathematical Reasoning," *Science* 210 (12 December 1990), 1262–1264. The study, finding that girls who took as many math classes as boys scored lower on the mathematical portions of standardized tests, was widely covered in the popular press. See discussion of response in Dorothy Nelkin, *Selling Science,* 2nd ed. (New York: W. H. Freeman, 1995).

13. *New York Times,* 7 December 1980.

14. *Time,* 15 December 1980.

15. Pamela Weintraub, "The Brain: His and Hers," *Discover,* April 1981, 15–20.

16. Jagger, *Feminist Politics* (Lanham, MD: Rowman, 1994).

17. Elizabeth Davies, *Women's Intuition* (Berkeley: Celestial Arts, 1989), 15.

18. Jane and Robert Handly, *Why Women Worry* (New York: Fawcett Crest, 1990). Popular books on the brain have often focused on sex differences. Psychiatrist Richard Restak, who wrote the script for the 1992 television program "The Brain," wrote in 1979 that sex differences are "innate, biologically determined, and relatively resistant to change." The two sexes, he said, have distinct thought processes and use their brains in different ways: *The Brain: The Last Frontier* (New York: Doubleday, 1979).

19. Anne Moir and David Jessel, *Brain Sex: The Real Difference Between Men and Women* (New York: Carol Publishing Group, 1991).

20. Advertisement, *Mirabella,* April 1991.

21. Meme Black and Jed Springarn, "Mind Over Gender," *Elle,* March 1992, 158–162.

22. Sheila Moore and Roon Frost, *Little Boy Book* (New York: Ballantine, 1986).

23. See discussion in Carol Tavris, *The Mismeasure of Woman* (New York: Simon and Schuster, 1992), 124–125.

George Gilder, "The Case Against Women in Combat," *New York Times Magazine,* 28 January 1989, 44.

24. Letter to the editor from Jesse D. Sheinwald, *New York Times,* 2 March 1990.

25. Cited in Ellen Goodman, "Troubling Omens from the Year of the Woman," *Newsday,* 22 December 1992.

26. Camille Paglia, *Sex, Art, and American Culture* (New York: Vintage, 1992), 108.

27. Robert Bly, *Iron John: A Book About Men* (New York: Vintage Books, 1990).

28. Trip Gabriel, "In Touch with the Tool Belt Chromosome," *New York Times,* 22 September 1991.

29. Clarissa Pinkola Estes, *Women Who Run with the Wolves* (New York: Ballantine, 1992), 8 and book jacket copy.

30. Kathy Silberger, "Sheep in Wolves' Clothing," *Village Voice,* 8 June 1993, 15.

31. Cited by Emily Martin, "Body Narratives, Body Boundaries," in Lawrence Grossberg, Cary Nelson, and Paula Treichler, eds., *Cultural Studies* (New York, London: Routledge, 1992), 409–423, esp. 413–14.

32. Meredith Small, "Sperm," *Discover,* July 1991, 48–53.

33. Martin, "Body Narratives," 415.

34. J. Phillipe Rushton expanded his ideas in *Race, Evolution and Behavior* (Newark: Transaction Books, 1994). Also see Stephen Jay Gould, *The Mismeasure of Man* (New York: W. W. Norton, 1981). Gould explores earlier efforts to measure brain size, either by weighing brains after death or by filling skulls with lead shot. He shows that attempts to find measures by which brain size was the crucial variable have always failed.

35. P. S., "Is Vincent Sarich Part of a National Trend?" *Science* 251 (25 January 1991), 369.

36. Arthur Jensen, "How Much Can We Boost IQ and Scholastic Achievement?" *Harvard Educational Review,* Winter 1969, 1–23. (The *Harvard Educational Review,* a student-edited publication, reaches about 12,000 readers.) See also Jensen, *Bias in Mental Testing* (New York: Free Press, 1979).

37. Jensen, cited in *Newsweek*, May 10, 1971.

38. "Can Negroes Learn the Way Whites Do?" *U.S. News and World Report,* 10 March 1969. This and the following citations on the controversy of the 1960s and 1970s were gathered by a Cornell University graduate student, David Livesay, "The Race/IQ Controversy in the Popular Press," 6 May 1983 (seminar paper).

39. "Is Intelligence Racial?" *Newsweek,* 10 May 1971.

40. Richard J. Herrnstein, "IQ," *Atlantic Monthly,* September 1971, 43–58.

41. "Creeping Meritocracy," *Newsweek,* 23 August 1971, 57.

42. Richard J. Herrnstein, "In Defense of Intelligence Tests," *Commentary,* February 1982, 40–50.

43. Richard J. Herrnstein, "IQ Testing and the Media," *Atlantic Monthly,* August 1982, 68–74.

44. Mara Snyderman and Stanley Rothman, *The IQ Controversy: The Media and Public Policy* (New Brunswick, NJ: Transaction Books, 1988).

45. Cited in Micaela di Leonardo, "White Lies, Black Myths," *Village Voice,* 22 September 1992, 31.

46. George F. Will, "Nature and the Male Sex," *Newsweek,* 17 June 1991, 70.

47. Warren Leary, "Struggle Continues over Remarks by Mental Health Officials," *New York Times,* 8 March 1992.

48. Frederick Goodwin, "Conduct Disorder as a Precursor to Adult Violence and Substance Abuse: Can the Progression Be Halted?" American Psychiatric Association Annual Convention, Washington, DC, May 1992.

49. Gay Floodedited and Sarah Ballard, "An Oddsmaker's Odd Views," *Sports Illustrated,* 25 January 1988, 7.

50. Tim Kurkjian, "West Al," *Sports Illustrated,* 16 April 1990, 66.

51. Reggie Jackson, "We Have a Serious Problem That Isn't Going Away," *Sports Illustrated,* 11 May 1987, 40.

52. "Nightline," 15 April 1987.

53. Paul Selvin, "The Raging Bull at Berkeley," *Science* 251 (25 January 1991), 368–371.

54. David Layzer, "Affirmative Action Is at Least on the Right Track," *New York Times,* 23 June 1990.

55. Susan Chira, "Minority Student Bias in Higher Education Quest," *New York Times,* 4 August 1992.

56. Pooley, "Dr. J," 34.

57. James Traub, "Professor Whiff," *Village Voice,* 1 October 1991.

58. For a history of the definition of race, see F. James Davis, *Who Is Black? One Nation's Definition* (University Park: Pennsylvania State University Press, 1991) and Lawrence Wright, "One Drop of Blood," *The New Yorker,* 25 July 1994, 46–59.

59. Spike Lee, *Mo' Better Blues,* Universal Pictures, 1990.

60. Toni Morrison, *Tar Baby* (New York: New American Library, 1981).

61. Toni Morrison, *Song of Solomon* (New York: New American Library, 1977), 155.

62. Michael Bradley, *Iceman Inheritance: Prehistoric Sources of Western Man's Racism, Sexism and Aggression* (New York: Kayode Press, 1991).

63. D. Hamer et al., "Androgen Involvement in Homosexuality," *American Journal of Human Genetics* 53 (1993), 844–852. Also see Robert Pool, "Evidence for a Homosexuality Gene," *Science* 261 (16 July 1993), 221–291.

64. David F. Greenberg, *The Construction of Homosexuality* (Chicago: University of Chicago Press, 1988), 1–3, 14, 18.

65. Ron Bayer, *Homosexuality in American Psychiatry* (New York: Basic Books, 1981).

66. Randy Shilts, *And the Band Played On* (New York: St. Martin's Press, 1987).

67. David Gelman, "Born or Bred," *Newsweek*, 24 February 1992, 48–53.

68. This was based on a report of a study published in the *Archives of General Psychiatry* that focused on 108 lesbians with identical and nonidentical twin sisters, plus 32 lesbians with adoptive sisters. The study results suggested that sexual preference depended on biology. Identical twins were much more likely to both be lesbians than were fraternal twins or sisters with no genetic relationship. "Most lesbians feel they were born gay," the report said. *Newsweek*, 22 March 1993, 53.

69. Quoted in Karen De Witt, "Quayle Contends Homosexuality Is a Matter of Choice, not Biology," *New York Times*, 14 September 1992.

70. The play, by Jonathan Tollins, starred Jennifer Gray.

71. "20/20," 24 April 1992.

72. See letters to the editor, *New York Times*, 27 July 1993.

73. Janet E. Hally, "Biological Causation of Homosexuality and Constitutional Rights," public lecture, New York University Law School, 11 October 1993.

74. *National Enquirer*, 10 August 1993.

75. Ron Wilson, "Study Raises Issue of Biological Basis for Homosexuality," *Wall Street Journal*, 30 August 1991.

76. Speech at a symposium on "The Homosexual Brain," City College, New York, 9 December 1991.

77. "Dear Abby" column, "Genes Are Key in Sexual Orientation," *Delaware County Daily Times*, 21 January 1992.

78. Marvin Harris, "Referential Ambiguity in the Calculus of Brazilian Racial Identity," in Norman Whitten Jr. and John Szwed, eds., *Africo-American Anthropology* (New York: Free Press, 1970), 75–86.

79. Thomas Laqueur, *Making Sex: Body and Gender from the Greeks to Freud* (Cambridge: Harvard University Press, 1990), 25–27.

80. Ibid., 201–227.

81. Anne Fausto-Sterling, "The Five Sexes: Why Male and Female Are Not Enough," *The Sciences*, March/April 1993, 20–25.

82. We are indebted to Sheila Murnaghan, who in her talk "Was Sex Different in the Ancient World?" (University of Pennsylvania, 17 February 1994) brought to our attention the extensive recent literature on sex and sexuality in ancient Greece. See also David M. Halperin, John J. Winkler, and Froma Zeitlin, eds., *Before Sexuality* (Princeton: Princeton University Press, 1990).

CHAPTER 7

1. Dorothy Nelkin and Sander Gilman, "Placing Blame for Infectious Disease," *Social Research*, Autumn 1988, 361–368.

2. Mary Douglas, *Natural Symbols* (New York: Penguin, 1970), 138.

3. Marion Smiley, *Moral Responsibility and the Boundaries of Community* (Chicago: University of Chicago Press, 1992), 13.

4. Shelby Steele, *The Content of Our Character* (New York: Harper and Row, 1991).

5. Physicians, for example, have been sued in a series of "wrongful life" cases for failure to use available diagnostic technologies.

6. Charles Murray, *Losing Ground: American Social Policy, 1950–1980* (New York: Basic Books, 1984). Murray's thesis

has remained highly visible in the media. *The Expanded Academic Index* listed, between January 1989 and August 1992, fifteen articles by Murray, ten articles that refer to his work, and fifteen reviews of his books.

7. He laid this position out again in a *New York Times* op-ed piece in 1992: "Stop Favoring Unwed Mothers," 16 January.

8. Thomas Byrne Edsall with Mary Edsall, *Chain Reaction* (New York: W. W. Norton, 1991).

9. Lawrence Mead, *The New Politics of Poverty* (New York: Basic Books, 1992). For a similar analysis, see Charles Sykes, *A Nation of Victims* (New York: St. Martin's Press, 1992).

10. Andrew Hacker, *Two Nations: Black and White, Separate, Hostile, Unequal* (New York: Scribners, 1992).

11. Cited in John Kuhn Bleimaier, letter to the editor, *New York Times*, 8 October 1981.

12. Daniel Koshland, "Elephants, Monstrosities, and the Law," *Science* 255:5046 (14 February 1992), 777.

13. James Q. Wilson, *The Moral Sense* (New York: Free Press, 1993).

14. David Garland, "Of Crime and Criminals," in Mike Maguire, Rod Morgan, and Robert Reiner, *Oxford Handbook of Criminology* (Oxford: Oxford University Press, 1995).

15. Harris B. Peck, "Are the parents the delinquents?" *New York Times Magazine*, 12 September 1954, 11.

16. Wertham, "Why Do They Commit Murder?" *New York Times Magazine*, 8 August 1954, 8.

17. C. Ray Jeffery, *Criminology: An Interdisciplinary Approach* (Englewood Cliffs, NJ: Prentice Hall, 1990), 184.

18. N. D. Volkow and L. R. Tancredi, "Positron Emission Tomography: A Technology Assessment," *International Journal of Technology Assessment and Health Care* 2 (1986), 577–594; "Market for DNA Probe Test for Genetic Diseases,"

Genetic Technology News, November 1986, 6–7; see also U.S. Congress, Office of Technology Assessment, *Medical Testing* (Washington, DC: USGPO, 1988), 132–143.

19. Quoted in Evelyn Nieves, "Life Sentence for Youth in Strangling," *New York Times,* 20 November 1993.

20. Berkeley Breathed, "Outland," *Washington Post,* 31 October 1993.

21. Lawrence Taylor, *Born to Crime* (Westport, CT: Greenwood Press, 1984).

22. Frederic Goodwin, "Conduct Disorder as a Precursor to Adult Violence and Substance Abuse," speech to the American Psychiatric Association Annual Convention, Washington, D.C., May 1992.

23. Speech by political scientist Ronald Walters to the National Medical Association, reported in *JAMA* 270:11 (15 September 1993), 1283.

24. Rachel Rubin, "Who Comes First?" *Parents,* May 1954, 34.

25. Robert B. McCall, "Good Parents Make Good Children or Do They?" *Parents,* June 1986, 108.

26. John Leo and Elizabeth Taylor, "Exploring the Traits of Twins," *Time,* 12 January 1987, 63.

27. Parental advice books and magazines offer insight into popular sentiments, for they are widely read. But it is difficult to assess how strictly their advice is followed. Readers bring to a text their own interpretations, and actual child-rearing practices may reflect personal experiences (parental models) more than parenting books. Letters to child advice authors such as Dr. Benjamin Spock reveal the tensions between reader and text. Many writers describe home situations that do not live up to the Spockian ideal and, while asking Spock's advice, suggest its irrelevance to their particular situation. See Michael Sulman, "The Humanization of the American Child: Benjamin Spock as the Popularizer of Psy-

choanalytic Thought," *Journal of the History of the Behavioral Sciences* 9:3 (July 1973).

28. Spock, 1945, vi, cited in Sulman, "Humanization," 258–265, 258. For reviews of Spock's changing ideas, see Christina Hardyment, *Dreambabies: Three Centuries of Good Advice on Child Care* (New York: Harper and Row, 1983), 267–285 and Nancy Pottishman, "Mother, The Invention of Necessity: Dr. Benjamin Spock's *Baby and Child Care,*" in N. Ray Hiner and Joseph M. Haweds, eds., *Growing Up in America: Children in Historical Perspective* (Urbana and Chicago: University of Illinois Press, 1985).

29. Irving Harris, *Emotional Blocks to Learning: A Study of the Reasons for Failure in School* (Glencoe, IL: Free Press, 1962).

30. Sindonie M. Gruenberg and Hilda Sidney Krech, "Are Parents Always to Blame?" reprinted in *Child and Family*, April 1962, 52–56.

31. Stella Chess, M.D., and Alexander Thomas, M.D., *Know Your Child* (New York: Basic Books, 1987), jacket copy.

32. Thomas Clavin, "What Did You Expect?" *Child*, December/January 1993, 61–65.

33. Dr. Benjamin Spock with Michael B. Rothenberg, *Dr. Spock's Baby and Child Care* (New York: Pocket Books, 1985), 22.

34. Carin Rubinstein, "Overload," *Parenting*, October 1991, 84.

35. Cited in Clavin, "What Did You Expect?" 65.

36. Ann E. LaForge, "Born Brainy," *Child*, November 1991, 88–90.

37. Julius Segal, Ph.D., and Zelda Segal, "Why Kids Are the Way They Are," *Parents*, August 1991, 80.

38. Nancy Shute, "How Healthy Is Your Family Tree?" *Hippocrates*, January/February 1988, 88–89.

39. Randi Londer, "Cracking the Code," *Child*, January/February 1988, 44–46.

40. Stanley Greenspan, *First Feelings: Milestones in the Emotional Development of Your Baby and Child* (New York: Viking, 1985).

41. Jerome Kagan, J. S. Reznick, and N. Shidman, "Biological Basis of Childhood Shyness," *Science* 240 (8 April 1988), 167–171.

42. Phil Donahue, "The Human Animal—Nature vs Nurture," Discovery Channel, 10 April 1992.

43. Valerie Adler, "Born Bashful," *American Health* 7 (October 1988), 90.

44. Joan Costello, "Shy Kids: Born or Made?" *Parents*, January 1983, 76.

45. Alvin P. Sanoff, "A Conversation with Jerome Kagan: 'Your Mother Did It to You' Is an Excuse Americans Overuse," *U.S. News and World Report*, 25 March 1985, 63.

46. Sandra Scarr, "Developmental Theories for the 1990s: Developmental and Individual Differences": Presidential address to the biennial meeting of the Society for Research in Child Development (Seattle, 20 April 1991), 4–5, 24.

47. "Childhood," 14 October 1991.

48. Joan Williams, "Gender Wars: Selfless Women in the Republic of Choice," *New York University Law Review* 66:6 (December 1991), 1605.

49. Erik Eckholm, "Learning If Infants Are Hurt When Mothers Go to Work," *New York Times*, 6 October 1992 (third and last in a series entitled "The Good Mother: Minding the Children").

50. Cathy Trost, "Type A Mothers Are More Likely to Have Children with Their Personality Traits," *Wall Street Journal*, 5 March 1992.

51. Chess and Thomas, *Know Your Child*, 6, 11.

52. Elaine Fantle-Shimberg, *Depression: What Families Should Know* (New York: Ballantine Books, 1991), 130, 190.

53. Terrence Monmaney, "When Manic Depression Is Part of the Family Legacy," *Newsweek*, 4 May 1987, 53.

54. Ron Howard, *Parenthood*, 1990.

55. Terence Rafferty, "Force of Nature," *The New Yorker*, 11 January 1993, 101–103.

56. Lars Hertzbert, "Blame and Causality," *Mind* 84 (1974), 550. Cited in Smiley, *Moral Responsibility*, 72.

57. Kenneth M. Grundfast, "Bring Back Berendzen," *Washington Post*, 4 May 1990. Following extensive therapy Berendzen, in public statements and a popular book, portrayed himself as a survivor and blamed his problems on child abuse. *New York Times*, 6 September 1993; and Richard Berendzen and Laura Palmer, *Come Here* (New York: Villard Books, 1993).

58. Charles Krauthammer, "Nature Made Me Do It," *Washington Post*, 11 May 1990.

59. American Psychiatric Association, *Quick Reference to the Diagnostic Criteria for DSM IV* (Washington, DC.: 1994), 246. See also Constance Holden, "Revising Psychiatric Diagnoses," *Science* 260 (June 1993), 1586–1587.

60. Berendzen and Palmer, *Come Here*.

61. See, for example, the popular books by Bernie Siegel, M.D., on the causes of cancer. The first of these was *Love, Medicine and Miracles* (New York: Harper and Row, 1986).

62. Jane E. Brody, "For Most Who Try to Lose Weight, Dieting Makes Things Worse," *New York Times*, 23 November 1992.

63. Jeannie Ralson, "30 Exercise Myths," *Allure*, June 1993.

64. Lionel Tiger, *Pursuit of Pleasure* (Boston: Little, Brown, 1992).

65. Molly O'Neill, "Eat, Drink and Be Merry May Be the Next Trend," *New York Times*, 2 January 1994.

66. Dan Hurley, "Arresting Delinquency," *Psychology Today*, March 1985, 63.

67. Lawrence A. Chavez, director, *Vivigen Genetic Repository* brochure, Santa Fe, New Mexico, 15 June 1991.

68. Shannon Brownlee and Joanne Silberner, "The Age of Genes," *U.S. News and World Report*, 4 November 1991, 64.

69. "Nobody's Perfect," *Mademoiselle*, January 1992, 78–79.

70. Opti-Genetics, Metabolic Nutrition, Inc., Miami, Florida.

71. Leo Buscaglia, *Living, Loving, and Learning* (New York: Holt, Rinehart and Winston, 1982). Cited in Irene Taviss Thomson, "Individualism and Conformity in the 1950s vs the 1980s," *Sociological Forum* 7:3 (September 1992), 502.

72. Some of this information about NAMI was graciously sent to us by Laura Lee Hall of the Office of Technology Assessment.

CHAPTER 8

1. Philip Kerr, *A Philosophical Investigation* (New York: Farrar, Straus & Giroux, 1992).

2. Mary Douglas, *How Institutions Think* (Syracuse, NY: Syracuse University Press, 1986), 63, 92.

3. Yaron Ezrahi, *The Descent of Icarus: Science and the Transformation of Contemporary Democracy* (Cambridge: Harvard University Press, 1990). For cases that illustrate this pattern, see Dorothy Nelkin, ed., *Controversy: The Politics of Technical Decisions* (Newbury Park, CA: Sage Publications, 1992).

4. The agenda-setting role of the media has been widely documented. See Gaye Tuchman, *Making News* (New York: Free Press, 1978).

5. For a fuller development of these ideas, see Dorothy Nelkin and Laurence Tancredi, *Dangerous Diagnostics: The*

Social Power of Biological Information, 2nd ed. (Chicago: University of Chicago Press, 1994).

6. See Melissa Arndt, "Severed Roots: The Sealed Adoption Records Controversy," *Northern Illinois University Law Review* 6 (1986), 103, 104–105.

7. Peggy C. Davis, "There Is a Book Out: An Analysis of Judicial Absorption of Legislative Facts," *Harvard Law Review* 100:7 (May 1987), 1539–1604.

8. See Elton B. Kilibanoff, "Genealogical Information in Adoption: The Adoptee's Quest and the Law," *Family Law Quarterly* 11 (1977), 185, 193.

9. Seth Mydans, "Surrogate Denied Custody of Child," *New York Times,* 23 October 1990.

10. Johnson *v.* Calvert, No. X 63 31 90 (Cal. Super. Ct. Oct. 22, 1990), at 8.

11. Ibid., at 11. See Anna J. *v.* Mark C., 286 Cal. Rptr. 369 (Ct. App. 1992) for appeal.

12. The influence of genetics is highlighted by the fact—not mentioned by the court, but evident in the illustrations in the press—that the Calverts were white and Johnson was black. Johnson appealed the decision, and in May 1993 the California Supreme Court denied her appeal on grounds that genes were "a powerful factor in human relationships." See Janet Dolgin, "Just a Gene: Judicial Assumptions about Parenthood," *UCLA Law Review* 40 (1993), 637–694.

13. Commonwealth *ex. rel.* Coburn *v.* Coburn, 558 A.2d at 548, 553 (Pa. Super. Ct. 1989). See also John Lawrence Hill, "What Does It Mean to Be a 'Parent'? The Claims of Biology as the Basis for Parental Rights," *New York University Law Review* 66 (1991), 353.

14. Judge J. Crillo, (concurring) in Commonwealth *ex. rel.* Coburn *v.* Coburn, at 554.

15. In a Vermont adoption case involving a nine-month-old baby, the court worked out a settlement between the biologi-

cal father and adoptive parents. See "Pact Reached in Vermont Adoption Case," *New York Times,* 23 August 1993.

16. Don Terry, "Storm rages in Chicago over revoked adoption," *New York Times,* 15 July 1994.

17. Anthony Depalma, "Custody Decision Dividing Experts," *New York Times,* 27 September 1992.

18. Rochelle Dreyfuss and Dorothy Nelkin, "The Jurisprudence of Genetics," *Vanderbilt Law Review* 45:2 (March 1992), 321.

19. Baker *v.* State Bar, 781 P. 2d 1344 (Cal. 1989).

20. In re Ewaniszyk, 788 P. 2d 690 (Cal. Ct. App. 1990).

21. "Law and Order," 17 November 1993.

22. David Garland, "Of Crime and Criminals," in Mike Maguire, Rod Morgan, and Roland Reiner, eds., *Oxford Handbook of Criminology* (New York: Oxford University Press, 1995).

23. Ibid. Garland observes that there is often a difference between the discourse and the reality, for treatment programs do persist.

24. Jake's Page, "Exposing the Criminal Mind," *Science* 84, September 1984, 84–85.

25. National Research Council, *Understanding and Preventing Violence* (Washington, DC: National Academy of Sciences Press, 1993), Ch. 1.

26. Roger Nemendez, M.D., letter to the editor, *JAMA,* 2 December 1992, 3070.

27. Lawrence Taylor, "The Genetic Defense," *Science Digest,* November 1982, 44.

28. Joel Keehn, "The Long Arm of the Gene," *The American Way,* 15 March 1992, 36–38. Citing William C. Thompson from the University of California, Irvine.

29. William Raspberry, "Criminal Types," *Washington Post,* 22 November 1985.

30. Cartoon by Nick Downes in *Science* 238 (9 November 1987), 772.

31. Dorothy Nelkin and Laurence Tancredi, *Dangerous Diagnostics.*

32. Benjamin Schatz, "The AIDS Insurance Crisis: Underwriting or Overreaching?" *Harvard Law Review* 100 (1987), 1782; Katherine Brokaw, "Genetic Screening in the Workplace and Employers' Liability," *Columbia Journal of Law and Social Problems* 23 (1990), 317, 327 cites cases in which genetic diseases have been excluded from coverage as "pre-existing conditions."

33. Larry Thompson, "The Price of Knowledge: Genetic Tests that Predict Conditions Became a Two-Edged Sword," *Washington Post,* 10 October 1989.

34. See U.S. Congress, Office of Technology and Assessment, "Genetic Monitoring and Screening in the Workplace" (Washington, DC: USGPO, 1990), 10–12.

35. Mark A. Rothstein, "Employee Selection Based on Susceptibility to Occupational Illness," *Michigan Law Review* (1983), 1379. Also see Elaine Draper, *Risky Business* (New York: Cambridge University Press, 1991).

36. Michael W. Miller, "In Vino Veritas: Gallo Scientists Search for Genes of Alcoholism," *Wall Street Journal,* 8 June 1994.

37. David Lauria, Mark Webb, Pamela McKenzie, M.D., and Randi Hagerman, M.D., "The Economic Impact of Fragile-X Syndrome on the State of Colorado," *Proceedings of the International Fragile-X Conference,* 1992, 393–405.

38. Laura Lesch, personal communication to authors, June 1994.

39. American Psychiatric Association, *Diagnostic Statistical Manual of Mental Disorders,* 4th ed. (Washington, DC: American Psychiatric Association, 1994). See discussion in Paul Cotton, "Psychiatrists Set to Approve DSM IV," *JAMA* 270 (7 July 1993), 13–15.

40. Robert Proctor, *Value-Free Science* (Cambridge: Harvard University Press, 1991), 5.

41. Nikolas Rose, *The Psychological Complex* (London: Routledge and Kegan Paul, 1985), 5.

42. C. R. Jeffery, *Attacks on the Insanity Defense: Biological Psychiatry and New Perspectives on Criminal Behavior* (Springfield, IL: Charles C. Thomas, 1985), 82.

43. Theodore Modis, *Predictions: Society's Telltale Signature Reveals the Past and Forecasts the Future* (New York: Simon and Schuster, 1992).

44. Laurence Tribe, "Trial by Mathematics: Precision and Ritual in the Legal Process," *Harvard Law Review* 84 (1971), 1329.

45. Paul R. Billings et al., "Discrimination as a Consequence of Genetic Testing," *American Journal of Human Genetics* 50 (1992), 476–482.

46. Abby Lippman argues that "[d]isorders and disabilities are not merely physiological or physical conditions with fixed contours. Rather, they are social products with variable shapes and distributions." "Prenatal Genetic Testing and Screening: Constructing Needs and Reinforcing Inequities," *American Journal of Law & Medicine* 17 (1991), 15, 17.

47. Louis Harris and Associates, "Genetic Testing and Gene Therapy: National Survey Findings," March of Dimes Birth Defects Foundation, September 1992.

CHAPTER 9

1. Leo Rost, letter to the editor, *New York Times*, 9 October 1991.

2. Daniel Seligman, *A Question of Intelligence: The IQ Debate in America* (New York: Carol Publishing Group, 1992).

3. Editorial, *Philadelphia Inquirer*, 12 December 1990.

4. Jeremy Rifkin, *Biosphere Politics* (New York: Harper Collins, 1991).

5. Daniel Kevles, "Controlling the Genetic Arsenal," *Wilson Quarterly*, Spring 1992, 68.

6. Marqui-Luisa Miringoff, *The Social Costs of Genetic Welfare* (New Brunswick, NJ: Rutgers University Press, 1991), 20.

7. Troy Duster, *Backdoor to Eugenics* (New York and London: Routledge, 1990), 112–129. Duster argues particularly that the idea that a "defective fetus" can be eliminated has penetrated the collective conscience, 113.

8. Carl N. Degler, *In Search of Human Nature* (New York: Oxford University Press, 1991).

9. Richard John Neuhaus, "The Return of Eugenics," *Commentary*, April 1988, 1–28.

10. This phrase was used in Nazi Germany to describe the program to kill mentally retarded, physically ill, and mentally ill Germans including newborns with birth defects. Robert Proctor, *Racial Hygiene; Medicine Under the Nazis* (Cambridge: Harvard University Press, 1988).

11. For other perspectives, see Diane Paul, "Eugenic Anxieties, Social Realities, and Political Choices," *Research* 59 (Fall 1992), 664–683. See also Evelyn Fox Keller, "Nature, Nurture, and the Human Genome Project," in Daniel Kevles and Hood (eds.), *The Code of Codes* (Cambridge: Harvard University Press, 1992), 281–299.

12. Jeremy Gerard, "CBS Gives Rooney a 3-Month Suspension for Remarks," *New York Times*, 9 February 1990.

13. Edwin M. Yoder, "In Defense of Rooney," *Washington Post*, 20 February 1990.

14. Andrew Hacker, *Two Nations* (New York: Scribners, 1992).

15. This is, of course, hardly a new theme. In the 1960s, Nobelist William Shockley associated social problems with the reproductive practices of the poor: "Many of the large

improvident families with social problems simply have constitutional deficiencies that . . . can be passed from one generation to another." Society, he said, cannot afford to keep
reproducing defective people who "lack the brains and moral
sense needed for good citizenship." He called for efforts to
encourage reproduction among those with superior ability.
See William Shockley, "Is the Quality of the Population
Declining?" *U.S. News and World Report,* 22 November 1965,
68–69.

16. W. David Kubiak, "E Pluribus Yamato, The Culture of
Corporate Beings," *Whole Earth Review,* Winter 1990, 4–10.

17. Daniel Seligman, *A Question of Intelligence: The IQ
Debate in America,* 118–127, 187.

18. Seligman op. cit.

19. See Richard Herrnstein, "A Confederacy of Dunces,"
Newsweek, 22 May 1989, 80–81; and Richard Herrnstein and
Charles Murray, *The Bell Curve* (New York: Free Press, 1994).

20. Ben Wattenberg, *The Birth Dearth* (New York: Pharos
Books, 1987), 1, 127.

21. Julie Smith, *True Life Adventures* (New York: Time-
Warner Books, 1985), 150–151.

22. Dennis Chamberland, "Genetic Engineering: Promise
and Threat," *Christianity Today,* 7 February 1986, 20.

23. Adrienne Asch, "Can Aborting 'Imperfect' Children Be
Immoral?," in her *Real Moral Dilemmas* (New York: Christianity and Crisis, 1986), 317–321.

24. Steven A. Holmes, "Abortion Issue Divides Advocates for
Disabled," *New York Times,* 4 July 1991.

25. Richard Saltus, "Paradox in Gene Therapy Debate,"
Boston Globe, 6 February 1989.

26. *Disability Rag,* January/February 1994.

27. Laura Lee Flynn, remarks at a conference on "Understanding the Role of Genetic Factors in Mental Illness,"

organized by the Office of Technology Assessment of the U.S. Congress and the NIMH, 22 January 1993.

28. Jill Brooke Coiner, "Equal in the Eyes of Love" (profile of Bree Walker Lampley), *Redbook*, April 1992, 146–148.

29. Jane Norris, KFI-RADIO, San Diego, California, 22 July 1991.

30. Jill Brooke Coiner, "Equal in the Eyes of Love."

31. *Bree Walker et al. v. KFI Radio Station,* transcript: Complaint before the Federal Communications Commission, Washington, DC, 21 October 1991.

32. "F.C.C. Rebuffs Handicapped Television Anchor," *New York Times,* 15 February 1992.

33. Robert Bogdan, *Freak Show: Presenting Human Oddities for Amusement and Profit* (Chicago: University of Chicago Press, 1988); Erving Goffman, *Stigma: Notes on the Management of Spoiled Identity* (New York: Simon and Schuster, 1986).

34. "Northern Exposure," 4 November 1991.

35. Fay Weldon, *The Cloning of Joanna May* (New York: Viking, 1989), 44–45.

36. "Life Goes On," 1991–1993.

37. "Star Trek: The Next Generation," "The Masterpiece Society" (episode), 15 February 1992.

38. Ben Klassen, "Triage," *Racial Loyalty,* 67 (January 1991), 1–2.

39. "Cupid's Corner," *Racial Loyalty,* 66 (December 1990), 4.

40. *National Alliance,* Arlington, Va.

41. Ben Klassen, "Racial Socialism," *Racial Loyalty,* 67 (January 1991), 7.

42. From: *Germanische Leitheft,* I Jahgang, Heft 1, 1941. Freely translated and edited in Eric Gustavson, "Europe

Becomes Healthy Again," *The New Order,* January/February 1991, 4.

43. Mari J. Matsuda, "Public Response to Racist Speech: Considering the Victim's Story," *Michigan Law Review* 87 (August 1989), 2320–2381. See also, for documentation of escalating violence of hate groups, Klanwatch Intelligence Report, February 1989; literature from the Anti-Defamation League of B'nai Brith; and Southern Poverty Law Center, *The Ku Klux Klan?: A History of Racism and Violence,* 3rd ed. (Montgomery, AL: Southern Poverty Law Center, 1988).

44. "Klanwatch Reports Record Rise in Hate Groups," *SPLC Reporter* 21:2 (April 1992), 1.

45. A 1993 Roper survey seemed to show that 22 percent of Americans had doubts about whether the Holocaust ever occurred, but a reassessment of that survey in 1994 suggested that this result was a consequence of a question that was poorly worded and easy to misunderstand. Michiko Kakutani, "Where History Is a Casualty," *New York Times,* 30 April 1993.

46. Julia Reed, "His Brilliant Career," *New York Review of Books,* 9 April 1992, 20.

47. Frances Frank Marcus, "White-Supremacist Group Fills a Corner in Duke," *New York Times,* 14 November 1991.

48. Robin Toner, "Duke Takes His Anger into 1992 Race," *New York Times,* 5 December 1991.

49. Anna Quindlen, "(Same Old) New Duke," *New York Times,* 13 November 1991.

50. Sharon L. Camp, Ph.D., "Population Pressure, Poverty and the Environment," a lecture delivered at the conference on "Endangered Earth: An Evolutionary Perspective," University of California, Los Angeles, 17 January 1990. Reprint (Washington, DC: Population Crisis Committee, 1989), 4, 23.

51. "Population Pressures Abroad and Immigration Pressure at Home" (Washington, DC: Population Crisis Committee, 1989).

52. Bernard Berelson, "The Great Debate on Population Policy: An Instructive Entertainment," *International Family Planning Perspectives* 16:4 (December 1990), 126–138.

53. David Berreby, "The Numbers Game," *Discover,* April 1990, 42–49.

54. Paul Ehrlich, cover letter, in Zero Population Growth mailing, 1991.

55. Robert H. Nelson, "Tom Hayden, meet Adam Smith and Thomas Aquinas," *Forbes,* 29 October 1990, 94–96.

56. Christopher Manes, "Why I Am a Misanthrope," *Earth First,* 21 December 1990, 28–29.

57. "Live Well or Diet," *Earth First,* 21 December 1990.

58. Feral Ape, "Dear Shit for Brains," *Earth First,* 20 March 1991, 4.

59. Theodore Roszak, "Green Guilt and Ecological Oversell," *New York Times,* 9 June 1992.

60. "Count Down to Extinction," *Megadeth* (album), 1992.

61. Francine Prose, "Would You Buy a Used Car from This Family?" *New York Times Book Review,* 12 January 1992.

62. "X-Cutioner's Song—Part I" in *Uncanny X-Men* 1:294, Marvel Comics, November 1992.

63. Hank Kanalz, Paul Pelletier, and Ken Brand, *Ex-Mutants,* Malibu Comics, March 1993.

64. *Extinction Agenda,* Marvel Comics, January 1, 1991.

65. Stan Lee, *Official Handbook of the Marvel Universe,* Vol. 8 (New York: Marvel Comics, 1987), 94–96.

66. (William) Olaf Stapledon, *Last and First Men* (London: Methuen, 1930; reprint, Los Angeles: Jeremy P. Tarcher, 1988), 81.

67. David Brin, *The Uplift War* (New York: Bantam Spectra, 1987), 631.

68. Nancy Kress, *Beggars in Spain* (New York: Avonova/Morrow, 1993).

69. Octavia Butler, *Dawn* (New York: Warner Books, 1987).

70. Arthur C. Clarke, *The Garden of Rama* (New York: Bantam, 1991).

71. "Island City," 4 March 1994.

72. Herrnstein and Murray, op. cit. and Seymour W. Itzkoff, *The Decline of Intelligence in America* (Westport, CT: Praeger, 1994). Other books on the genetic basis of behavior, all appearing at the end of 1994, include Jerome Kagan, *Galen's Prophecy: Temperament in Human Nature* (New York: Basic Books, 1994); Robert Wright, *The Moral Animal* (New York: Pantheon, 1994); and J. Phillippe Rushton, *Race, Evolution and Behavior* (New Brunswick, NJ: Transaction Books, 1994). Some of this work has been supported by the Pioneer Fund, a 57-year-old foundation concerned with promoting eugenics. See Barry Mehler, "In Genes We Trust," *Reform Judaism*, Winter 1994, 23:2, 10–79 on the Pioneer Fund.

73. Herrnstein and Murray, op. cit.

74. Ann Landers column, *Newsday*, 28 June 1991.

75. See discussion in Ronald Bayer, "AIDS and the Future of Reproductive Freedom," in Dorothy Nelkin, David Willis, and Scott Parris, eds., *A Disease of Society* (New York: Cambridge University Press, 1991), 191–215 (quotation is on page 193).

76. Editorial, "Poverty and Norplant," *Philadelphia Inquirer,* 12 December 1990.

77. Faye Wattleton, "Using Birth Control as Coercion," *Los Angeles Times,* 15 January 1991.

78. Cal Thomas, "A short step between extortion and coercion," *American Republic,* 30 January 1990.

79. Jonathan Alter, "One Well-Read Editorial," *Newsweek,* 31 December 1990.

80. Michael Lev, "Judge Is Firm on Forced Contraception, but Welcomes an Appeal," *New York Times*, 11 January 1991.

81. G. Skelton and D. Weinberg, "Most Support Norplant for Teens and Drug Addicts," *Los Angeles Times*, 27 May 1991.

82. Tamar Lewin, "A Plan to Pay Welfare Mothers for Birth Control," *New York Times*, 9 February 1991.

83. Julie Mertus and Simon Heller, "Norplant Meets the New Eugenicists," *Saint Louis University Public Law Review* 11 (1992), 359–383.

84. Randall Fasnacht, *Life Child: The Case for Licensing Parents* (Albany, NY: Life Force Institute, 1992).

85. "Not least among the requirements for this would be a far more thoroughgoing and widespread education of the public in biological and social essentials. And there would also have to be a very great improvement in the technical methods whereby the more important features of the genetic constitution may be judged." H. J. Muller, "Our Load of Mutations," *American Journal of Human Genetics* 2:2 (June 1950), 111–176.

86. Dr. Robert K. Graham, Repository for Germinal Choice, Escondido, California; interview with authors, 3 January 1992.

87. See discussion in Maureen McNeil, Ian Varcoe, and Steven Yearly, *The New Reproductive Technologies* (New York: St. Martin's Press, 1990).

88. "This Is What You Thought About . . . Breeding for Intelligence," *Glamour*, August 1980.

89. "Superkids? A Sperm Bank for Nobelists," *Time*, 10 March 1980.

90. Stanley N. Wellborn, "A Genetic Elite Taking Shape in the U.S.," *U.S. News and World Report*, 24 March 1980.

91. Lori B. Andrews, "Inside the Genius Farm," *Parents*, October 1980.

92. D. Keith Mano, "The Gimlet Eye: The Nobel Sperm Bank," *National Review,* 12 November 1982.

93. ESIG is part of MENSA. Based in Tennessee, it occasionally publishes *Eugenics Bulletin.*

94. Dr. Robert K. Graham, interview, 3 January 1992.

95. Paul Smith, 28 April 1992; telephone interview with authors. This bank has eight white donors who keep their sperm "on file," and according to Smith, most are professors from the California Institute of Technology.

96. Jean Bethke Elshtain, "The New Eugenics and Feminist Quandaries: Philosophical and Political Reflections," in Richard Neuhaus, ed., *Guaranteeing the Good Life* (New York: Eerdmans, 1990), 68–89.

97. David Lauria et al., "The Economic Impact of the Fragile-X Syndrome on the State of Colorado," International Fragile-X Conference Proceedings, 1992.

98. Rena Yount, "Pursuit of Excellence," in Damon Knight, ed., *The Clarion Awards, 1984* (New York: Doubleday, 1984).

99. According to Robert Proctor, *Racial Hygiene,* nurses and physicians in Nazi Germany continued to kill mental patients and others for months after the Allied occupation, and some parents said they wanted their handicapped children to be taken away by Nazi authorities. Indeed, the Nazis defended their eugenic policies by suggesting that people *choose* to accept their reproductive responsibilities. See Chapters 5 and 9.

CHAPTER 10

1. Sandra Scarr, "Three Cheers for Behavioral Genetics" (presidential address to the Behavioral Genetics Association), *Behavior Genetics* 17:3 (1987), 219–228. She writes: "Those were heady days, not unlike fighting the Nazis of an earlier era." (219)

2. Several years later, psychiatrist Richard Restak echoed Scarr's comments. He recalled how he was once "hooted down" by feminists objecting to his studies of sex differences in brain structure. Now, however, he finds that people are eager to hear more. Quoted in Meme Black, "Mind over Gender," *Elle*, March 1992, 161–162.

3. David Bloor, "Twenty Industrial Scientists," in Mary Douglas, ed., *Essays in the Sociology of Perception* (London: Routledge and Kegan Paul, 1982), 83.

4. Albert Rosenfeld, "The Medical Story of the Century," *Longevity*, May 1992, 42–53.

5. For discussion of this construct, see Jonathan Simon, "The Emergence of a Risk Society: Insurance Law and the State," *Socialist Review* 95 (September/October 1987), 63–89.

6. Dr. Martin E. P. Seligman, quoted in Molly O'Neil, "Eat, Drink and Be Merry May Be the Next Trend," *New York Times*, 2 January 1993. Some examples of recent advice books with this theme are Seligman's *What You Can Change and What You Can't Change* (New York: Knopf, 1993); and Jeffrey Harris, *Deadly Choices* (New York: Basic Books, 1993).

7. "Has Nature Overwhelmed Nurture?" *Nature* 366 (11 November 1993).

8. James Crow, "Eugenics: Must It Be a Dirty Word?" *Contemporary Psychology* 33 (January 1988), 9–12.

9. James Watson, interviewed in *Issues in Science and Technology*, Fall 1993, 44–45.

Sources

PAGE 7
Symbol from *Human Genome News*. U.S. Department of Energy and the National Institutes of Health

PAGE 17
Copyright © 1994 Time Inc. Reprinted by permission

PAGE 23
Drawing from Hooton, E.A. *Crime and the Man.* New York: Greenwood Press, 1968, p. 245

PAGE 28
Photograph from *Eugenics in Race and State*, Vol. 11, Scientific papers of the Second International Congress of Eugenics. Baltimore: Williams and Wilkins Co., 1923

PAGE 45
Copyright © 1983 Mark Evanier and Will Meugniot. Reprinted by permission

SOURCES

PAGE 48
"Genetic Code Certificate" copyright © Larry Miller. Reprinted by permission

PAGE 67
Genetic roots drawing copyright © by Bill Sanderson

PAGE 86
Courtesy of USA Cable. Reprinted by permission

PAGE 108
Drawing by D. Reilly; © 1991. The New Yorker Magazine, Inc. Reprinted by permission

PAGE 123
Copyright © 1994, *Newsweek*, Inc. All rights reserved. Reprinted by permission

PAGE 160
Copyright © 1987 by Nick Downes; from SCIENCE. Reprinted by permission

PAGE 166
Copyright © 1990 by The New York Times Company. Reprinted by permission

PAGE 183
Genetic testing drawing reprinted by permission of Ronald Searle.

Index

Addictive behavior, 91–94,
  100, 161–162
Adoption, 62, 66–72
  birth parent searches and,
    71–72, 152
  custody disputes in,
    74–76, 151–155
  identity and, 66–70
  opposition to, 71
  sibling separation in,
    75–76
  stigmatization of, 70–71
Adrenaleukodystrophy, 143
Advertising, 13, 97
African Americans. *See also*
    under Racial
  classification of, 117–118
  genetic determinism and,
    112–118
  racial purity of, 118–119
Afrocentrism, 103, 117
Aggression. *See also* Crime

gender and, 108–109,
    110–111
  genetic basis of, 83–91
  race and, 115
AIDS, 120–121
Alcoholics Anonymous,
    91–92
Alcoholism, 91–94, 100,
    161–162
Alien III, 156–157
Allen, Steve, 99
American Eugenics Society,
    27, 29, 35
American Psychiatric Asso-
    ciation, 120, 122, 135,
    144
Americans with Disabilities
    Act, 167
Animal Lover (Donaldson),
    55–56
Ardrey, Robert, 90
Artificial insemination, 65

Artificial insemination *(continued)*
 donor selection for, 97–98, 188–189
Asimov, Isaac, 96
Athletic ability, racial differences in, 116

Baby and Child Care (Spock), 137–138
Baby Jessica case, 75, 154
Baby M case, 1
Baby Richard case, 154
Baker, La Reine Helen, 22
Baker v. State Bar, 156
Baltimore, David, 9
Barrymore, Drew, 100
Barrymore, John, Jr., 100
Bartholet, Elizabeth, 70–71
Batteau, Allen, 12
Beggars in Spain (Kress), 184
Behavior, genetic basis of, 9–10
 and gender, 104–112
 and race, 112–118
Behavioral genetics, 9–10, 77–78, 80–101, 104–118
 media interest in, 81–82
 twin studies in, 9, 81–82
The Bell Curve (Herrnstein & Murray), 181–182, 185
Benbow, Camille, 107
Berendzen, Richard, 144
Biological predisposition. *See* Genetic predisposition
Biomimicry, 43

Birth control, 185–187
Blacks. *See* African Americans
Blade Runner, 43–44
Blame, 127. *See also* Responsibility
 avoidance of, 144–148
 parental, 137, 138, 141, 142
The Blood of the Fathers (Lydston), 21–22
Bloor, David, 194
Bly, Robert, 110–111
Bogdan, Robert, 176
Born to Crime (Taylor), 135
Bouchard, Thomas, 81–82, 114
Bradley, Michael, 118
Brain Sex: The Real Differences Between Men and Women (Moir & Jessel), 109
Breathed, Berkeley, 134–135
Brin, David, 183
Buchanan, Pat, 99, 179
Buck v. Bell, 24
Burbank, Luther, 19, 21
Burke, Chris, 177
Buscaglia, Leo, 14
Bush, George, 99, 158
Buss, David, 77–78
Butler, Octavio, 184
Bynum, Carolyn, 41

Camp, Sharon, 180
Campanis, Al, 116
Cavalli-Sforza, Luca, 52
Character traits, genetic basis of, 9–10, 77–78, 80–101. *See also*

Behavior; Behavioral genetics
childrearing and, 140–142
Chess, Stella, 138, 142
Childrearing. *See* Parents
Children
  advice literature to parents, 139–141
  disputed custody of, 74–76, 151–155
  as genetic commodities, 97–99, 191
  intelligence in, 138. *See also* Intelligence
  success of, parental responsibility for, 136–144
  temperament in, 138–141
Christianity Today, 54
Clarke, Arthur C., 184
The Cloning of Joanna May (Weldon), 177
Coburn v. Coburn, 153–154
Coleman, Lester L., 88
Coltrane, Ravi, 98
Comic books, 44–46, 182–183
Contraception, 185–187
Cook, Harry H., 19
Cook, Robin, 56
Crichton, Michael, 53–54
Crick, Francis, 174
Crime
  DNA fingerprinting in, 47, 48–49
  genetic predisposition to, 21–24, 83–89, 134–135, 144, 155–156

race and, 115
  responsibility for, 132–136, 155–159
  sentencing decisions, 156–159
"Criminal chromosomes," 134
Criminology, 133–134
Crow, James, 198
Culture. *See* Popular culture
Custody disputes, 74–76, 151–155
Cyberspace, 42–43
Cyborg, 43

Dack family, 26
Darwin, Charles, 3
Daughters of the American Revolution, 76
Davenport, Charles, 26
Davis, Elizabeth, 109
Davis, Peggy Cooper, 12
Dawkins, Richard, 53, 107
Dawn (Butler), 184
DeBoer, Jessica, 75, 154
"Deep Space Nine," 48–49
Degler, Carl, 170
Deviance. *See* Crime
DeVries, Hugo, 3
Diagnostic and Statistical Manual of Mental Disorders (DSM), 120, 144, 162–163
Differential reproduction, 171–174, 181–187
Disabilities. *See also* Genetic diseases
  genetic testing for, 162–163, 166–168

Disabilities *(continued)*
  prenatal, 174–178
  reproductive rights and,
    147, 174–178
Discrimination, genetic test-
    ing and, 165–168
Disney, Walt, 70
Divorce, genetic explanation
    of, 77–78
DNA. *See also* Gene(s)
  as boundary marker,
    41–49
  as historical text, 52–53
  celebrity, 49–51
  immortality of, 52–53
  social uses of, 3, 106,
    124–126, 193–199
  as source of identity,
    41–49
  spiritual aspects of, 2–3,
    39–57
"DNA family," 72–78
DNA fingerprinting, 47,
    48–49
DNAgents, 44
DNA testing. *See also*
    Genetic testing
  in criminal investigations,
    47, 48–49
  on television, 73–74
Dodd, Westly Allan, 87
Donahue, Phil, 84, 92
Donaldson, Stephen R., 55
Douglas, Mary, 128
Drug addiction, 91–94
Dugdale, Richard, 25
Duke, David, 179–180
Dundy, Elaine, 79–80
Duster, Troy, 170

Ecology, reproductive con-
    trol and, 180–181
Ectrodactyly, 175–176
Edsall, Mary, 131
Edsall, Thomas, 131
Ehrlich, Paul, 181
Einstein, Albert, 98–99
Eliade, Mircea, 41
Employees, genetic testing
    of, 160–161, 167
Entitlement programs, criti-
    cism of, 130–132,
    184–185
Environmental degradation,
    overpopulation and,
    180–181
Environment vs. heredity
    debate, 9–10, 94–95
Ernest Gallo Clinic and
    Research Center,
    161–162
Estes, Clarissa Pinkola,
    111
E.T., 38
Eugenics, 19–37
  as civic religion, 30–33
  contraception and,
    185–187
  criminality and, 21–24
  family studies in, 25–27
  germplasm in, 20, 30, 36,
    37
  in infertility research, 34
  in Nazi Germany, 33
  popularization of, 20–22
  in postwar era, 33–37,
    169–192
  reproductive control and,
    23–24, 174–178. *See*

also Reproductive
control
selective breeding and,
27–30, 170–174,
185–192
survival of fittest and,
178–184
utopian aspects of, 31–32
voluntary, 190–192
Eugenics competitions,
27–30
Evolution, eugenics and,
178–184
Evolutionary psychology,
77–78
Evolutionary synthesis, 39
Extinction, eugenics and,
178–184

Falwell, Jerry, 121
Family
instability of, 59–60,
77–78
"molecular," 58–78
pathological, 25–27
as social vs. genetic entity,
58–61, 72–78
TV portrayals of, 72–74
"Family Secret," 69
Family trees, 76–77
Fasnacht, Randall, 187
Fausto-Sterling, Anne, 125
Feminism, 103, 106
Fetal protection, 65
Fetal testing, for genetic
disorders, 174–178,
189
First Feelings (Greenspan),
140

Fleming, Anne Taylor, 61–62
Flick, Lawrence F., 20
Flynn, Laura, 175
Folger, Felicia, 20
Fox, Robin, 90
Fragile-X, testing for, 162
Frost, Roon, 109

Galen's Prophecy (Kagan),
140
The Garden of Rama
(Clarke), 184
Garland, David, 133
Gaye, Marvin, 98
Gay gene, 119–120. *See also*
Homosexuality
Gender differences, 104–112
social meaning of,
125–126
Gene(s). *See also* DNA
behavior and, 9–10, 77–78,
80–101, 104–118
conceptual derivation of,
3–4
cultural meaning of, 1–4,
12–18, 193–199
defined, 2, 3
gay, 119–120
power of, 6–7, 193–199
spiritual meaning of, 2–3,
39–42, 54–57
Genealogy, 76–77
Gene mapping, 5–6, 8–9,
57
Genetic determinism, 100–
101. *See also* Genetic
predisposition
limitations of, 196–199
race and, 112–118

Genetic determinism *(continued)*
social utility of, 100–101, 106, 124–126, 145–168, 196–199
Genetic diseases
guilt in, 143
insurers' interest in, 160
mapping of, 5–6
reproductive rights and, 147
screening for, 174–178, 189–192
Genetic engineering, 54–57
Genetic essentialism, 41–49, 57
adoption and, 68, 151–155
applications of, 149–168
in childrearing, 143–144
group differences and, 101–126
limitations of, 196–199
social utility of, 193–199
Genetic markers, 5, 6, 92–93
Genetic predisposition, 100–101
childrearing and, 136–144
correlation vs. cause and, 166
criminal behavior and, 134–136, 155–159
discrimination and, 165–168
risk prediction and, 159–168
social policy and, 100–101, 106, 124–126, 145–168, 196–199

testing for, 159–164. *See also* Genetic testing
Genetic responsibility, 190–192
Genetics, behavioral, 9–10, 77–78, 80–101, 104–118
Genetic testing
abuses of, 165–168
for crime prevention, 158–159
in criminal investigations, 47, 48–49
discrimination and, 166–168
of employees, 160–161, 167
prenatal, 174–178, 189
privacy issues in, 167–168
for risk prediction, 159–164
in schools, 162–164
Genome
images of, 6–8
mapping of, 5–9, 57
Gerbner, George, 67
Germplasm, 20–21, 30, 36, 37
Gilbert, Walter, 7
Gilder, George, 110
Gitlin, Todd, 12
Goldberg, Steven, 109
Goldman, Albert, 80
Goldstein, Jeffrey, 90
Goleman, Dan, 92–93
Goodwin, Frederick, 115, 135
Gould, Stephen Jay, 10
Graham, Robert K., 188

Green, Jeremy, 88
Greenberg, David, 120
Greenrage (Manes), 181
Greenspan, Alan, 128
Greenspan, Stanley, 140
Group differences, 101–
126
gender and, 104–112
race and, 112–118
sexual orientation and,
119–124
social meaning of,
124–126
Guilt. *See also* Blame
in genetic disease, 143
Guist, Allen, 107
Guttmacher, Alan F., 34

Hacker, Andrew, 131
Hall, Marianne Mele, 115
Halley, Janet, 122
Hamer, Dean, 119, 145
Handly, Jane, 109
Handly, Robert, 109
Haraway, Donna, 43
Harris, Irving, 138
Harris, Marvin, 124
Hate groups, 178–180
Hawking, Stephen, 39
Health insurance, for
genetic diseases, 160
Heller, Simon, 187
Heredity vs. environment
debate, 9–10, 94–95
social utility of, 106,
193–199
Hermaphrodites, 125
Herrnstein, Richard, 114,
173, 184–185

Hill folk, 25–26
Hirschfeld, Abraham, 100
Holmes, Oliver Wendell,
24
Holtzman, Elizabeth, 96
Homophobia, 122
Homosexuality, 103,
119–123
social meaning of, 120,
125
Hrdy, Sarah Blaffer, 105
Human Genome Diversity
Project, 52–53
Human Genome News, 6–7
Human Genome Project,
5–9
criticism of, 57
promotion of, 6–9
Hypothalamus, in homosex-
uals, 119, 121

The Iceman Inheritance
(Bradley), 118
Identity
adoption and, 66–72
DNA as source of, 41–49
gender, 104–112
Immortality, of DNA, 52–53
The Imperial Animal (Tiger
& Fox), 90
Inequality, legitimization of,
106–109
*The Inevitability of Patri-
archy* (Goldberg), 109
Infertility, 61–66
eugenics and, 34
social meaning of, 59–53
treatment for, 63–66,
187–189

In re Ewaniszyk, 156
Insurance, health, 160
Intelligence
  environmental stimula-
    tion and, 139
  genetic basis of, 114
  race and, 112–118,
    181–182
  reproductive rates and,
    172–174, 181–182,
    184–187
  selective breeding for,
    97–98, 188–189
  wealth and, 172–173
In vitro fertilization, 64, 65,
    189
Iron John (Bly), 110–111
Ishmaelite family, 25

Jackson, Reggie, 116
Jacobs, Patricia, 83
Jagger, Alison, 109
Jeffries, Alec, 47
Jeffries, Leonard, 103, 117
Jellinek, E. M., 91
Jensen, Arthur, 81, 113, 114,
    193
Jensen, Jim, 175
Jessel, David, 109
Jimmy the Greek, 116
Johannsen, Wilhelm, 3
Johnson, Mark, 12
Johnson, Roswell Hill, 19,
    37
Johnson v. Calvert, 152–
    153
Jordan, David Starr, 21
Jordan, Harvey Ernest, 21
Joyce, James, 96

Jukes family, 25–26
Jurassic Park (Crichton),
    53–54

Kagan, Jerome, 140
Kallikak family, 25
Kaplan, Deborah, 174
Kellogg, John, 21
Kerr, Philip, 149–150, 165
Kevles, Daniel, 170
Kingsley, Gregory, 154–155
Kinship, 59
Kirchner, Otakar, 154
Kite, Elizabeth, 25
Know Your Child (Chess &
    Thomas), 138, 142
Konner, Melvin, 46, 88
Koppel, Ted, 121
Koshland, Daniel, 88, 133
Kress, Nancy, 184
Kubiak, David, 172
Kundera, Milan, 13

Lakoff, George, 12
Lampley, Bree Walker,
    175–176
Landers, Ann, 185
Laqueur, Thomas, 125
Last and First Men (Staple-
    don), 183
Learning disabilities,
    162–164
Lederman, Leon, 39
Lee, Spike, 118
Lesbianism, 119–123
LeVay, Simon, 103, 119, 121,
    122, 145
Levin, Michael, 116
Levy, Jerre, 104

Lewis, Michael, 105
Life Child (Fasnacht), 187
"Life Goes On," 177
Lifton, Betty Jean, 66
Lincoln, Abraham, 51–52
Lipsitz, George, 72
Little Boy Book (Moore & Frost), 109
Look Who's Talking, 61
Lorenz, Konrad, 90
Lorenzo's Oil, 143
Losing Ground (Murray), 130–131
The Lost Father (Simpson), 68
Lydston, G. Frank, 21

Maccoby, Eleanor, 105
Macy Conference, 35
Male behavior, determinants of, 104–112
Manes, Christopher, 181
Mapping, gene, 5–6, 8–9, 57
Marfan syndrome, 51–52
Maric, Miliva, 98–99
Markers, genetic, 5, 6, 92–93
Marriage, genetic factors in, 77–78
Martin, Emily, 112
Mass culture. *See* Popular culture
Mathematical aptitude and heredity, 107–108
Mays, Kimberly, 74–75, 155
McKusick, Victor, 35
Mead, Lawrence M., 131
Media. *See* Popular culture; Television

Medical genetics, 195–196
Medicalization
    of alcoholism, 91–94
    of homosexuality, 120
Mednick, Sarnoff, 88
Melanin, 103, 117
Mental illness, reproductive rights and, 147
Mertus, Julie, 187
Metaphor(s), 7–8, 12–16
Milgram, Stanley, 83
Miller, Larry, 47
Miringoff, Marqui Luisa, 170
Mo Better Blues, 118
Moir, Ann, 109
Molecular family, 58–78
Moore, Sheila, 109
Morality, genetic basis of, 133
Morgan, Elaine, 106
Morris, Desmond, 90
Morrison Toni, 118
Motherhood Deferred (Fleming), 61–62
Mothers, 136–144. *See also* Parents
    blaming of, 137, 138, 141, 142
    working, 141–142
Muller, Herman J., 187–188, 191
Mullis, Kary, 49, 50, 51
Murray, Charles, 128, 130–131, 181, 184
Mutation (Cook), 56

The Naked Ape (Morris), 90
Nam family, 25, 26–27

National Alliance for the
Mentally Ill (NAMI),
147
National Institutes of Health
(NIH), 5
Nature's Thumbprint
(Neubauer), 139
Nature vs. nurture debate,
10, 94–95
Nay, Virginia June, 28
Nearing, Scott, 20
Neoconservative theories,
130–132, 178–180
Neo-Nazis, 178–180
Neubauer, Peter, 139
Neuhaus, Richard, 170
New Catholic Encyclopedia,
50
The New Politics of Policy
(Mead), 131
Nobel Sperm Bank, 188
Norplant, 185–187
Norris, Jane, 175

Obesity and genes, 93, 145
Obsessive compulsive disor-
der, 93
On Aggression (Lorenz), 90
Osborne, Frederick, 35
Overpopulation, 180–181

Paglia, Camille, 48, 85, 110
Pangenesis, 3
Parenthood, 142–143
Parents
blaming of, 137, 138, 141,
142
custody disputes and,
74–76, 151–155

reproductive rights of. *See*
Reproductive control
responsibility of, 136–144
Perot, Ross, 99
Personality traits. *See*
Behavior; Character
traits
genetic basis of, 9–10,
77–78, 80–101
A Philosophical Investigation
(Kerr), 149–150, 165
Plain Truth, 55
Polgar, Judit, 94–95
Polymerase chain reaction
(PCR), 49
Popenoe, Paul, 19, 37
The Population Bomb, 181
Population control, 180–
181
Population Crisis Commit-
tee, 180–181
Poverty
reproductive rights and,
171–174, 184–187
welfare and, 130–132,
184–185
Predisposition, biological.
See Genetic predispo-
sition
Pregnancy. *See also* Infertility; Reproductive con-
trol
fetal environment in, 65
social meaning of, 64
surrogate, 64–66, 152–
153
Prenatal testing, 174–178,
189
Presley, Elvis, 79–80

Prime Time Families (Taylor), 72
Privacy, genetic testing and, 167–168
Prodigy (Stewart), 56
"Pursuit of Excellence" (Yount), 190
The Pursuit of Pleasure (Tiger), 145

Quayle, Dan, 121, 128
Quindlen, Anna, 99

Race Betterment Conference, 27–30
Racial differences, 112–118, 181–182
  social meaning of, 124–126
Racial purity, 118–119, 178–180
Rafter, Nicole Hahn, 24
Rapp, Rayna, 12
Reagan, Nancy, 92, 99
Reagan, Ronald, 132
Recombinant DNA research, 54–57
Redman, Josh, 98
Reichler, Max, 20
Reitman, Ivan, 46
Reproductive control, 174–192
  disabilities and, 147, 174–178
  eugenic. *See* Eugenics
  genetic responsibility and, 190–192
  population control and, 180–181

poverty and, 171–174, 184–187
  racism and, 178–180
  voluntary, 190–192
Reproductive rates (*See* Differential reproduction)
Resolve, 63
Responsibility, 127–148
  avoidance of, 144–148
  in criminal justice, 132–136
  genetic, 190–192
  genetic determinism and, 100–101, 106, 196–199. *See also* Genetic predisposition
  neoconservative theories of, 130–132
  parental, 136–144
Rifkin, Jeremy, 56, 170
Risk prediction and avoidance, 159–168
Rooney, Andy, 171
Rose, Nikolas, 164
Rothman, Stanley, 114
Rush, Benjamin, 91
Rushton, J. Phillipe, 112–113

Sadler, William S., 22
Safire, William, 98–99
Sato, Yuko, 98
Scarr, Sandra, 110, 140, 141, 193–194
Schlafly, Phyllis, 107
Schmidt, Daniel, 75
Schneider, David, 59

Science
    objectivity in, 10–11,
        164
    promotion of, 6–11
Science fiction, genetic
    essentialism in,
    43–46, 48–49
Scott, Ridley, 43
Screening, for genetic dis-
    eases, 174–178,
    189–192
Selective breeding, 27–30,
    171–174, 187–192.
    See also Eugenics
contraception and,
    185–187
The Selfish Gene (Dawkins),
    53, 107
Seligman, Daniel, 172
Seligman, Martin, 145
Seville Statement on
    Violence, 90–91
Sex-based differences,
    104–112
    social meaning of,
        125–126
Shannon, T. W., 23
Shilts, Randy, 122
Shneour, Elie, 99
Shyness, 140–141
Siblings, adoptive separa-
    tion of, 75–76
Simpson, Mona, 68
Skinner, B. F., 83
Smiley, Marion, 128
Smith, Paul, 188
Smoot, George, 39
Snyderman, Mara, 114

Social problems
    medicalization of, 91–94,
        120
    responsibility for. *See*
        Responsibility
Society for the Study of
    Social Biology, 35
Song of Solomon (Morri-
    son), 118
Soul, DNA and, 2–3, 40–42
Speck, Richard, 84
Sperm, as invaders, 111–
    112
Sperm banks, 97–98,
    188–189
Spielberg, Stephen, 38, 53
Spock, Benjamin, 137–
    139
Sports, racial differences
    and, 116
Stanley, Julian, 107
Stapledon, Olaf, 183
Starr, Ringo, 98
"Star Trek: The Next Gener-
    ation," 177
Steele, Shelby, 128
Stereotyping, sexual,
    104–112
Sterilization, eugenics and,
    23–24
Stettler, Charlie, 96
Stewart, Michael, 56
Stokes, William Earl Dodge,
    21
Strathern, Marilyn, 12, 59
Students, genetic testing of,
    162–164
Surrogacy, 64–66

custody disputes in, 152–153
Swinburne, Richard, 40

"Tainted Blood," 86–87
Talent, genetic basis of, 94–101
Tar Baby (Morrison), 118
Taylor, Ella, 72, 73
Taylor, Lawrence, 135
Television
    family depiction on, 72–74
    power of, 12–13
Temperament, child development and, 138–141
The Territorial Imperative (Ardrey), 90
Testosterone, aggression and, 108
Thomas, Alexander, 138, 142
Tiger, Lionel, 90, 145
Time Passages (Lipsitz), 72
Tomorrow's Children, 32–33
Turkle, Sherry, 16
Turlington, Christie, 96
Twigg, Arline, 74–75
Twins, 46
Twin studies, 9, 81
Two Nations (Hacker), 131

Uplift War (Brin), 183–184
U.S. Department of Energy (DOE), 5

Van Buren, Abigail, 123
Victimism, 132, 135

Violence. See also Crime
    genetic basis of, 83–91, 134–135
    race and, 115
Violence Prevention Initiative, 158
Vivigen genetic repository, 146
Vos Savant, Marilyn, 99

Wahlsten, Douglas, 10
Wallace, Bruce, 8
Walters, Barbara, 122
Wanger, James, 134
War
    gender differences and, 110
    genetic basis of, 89–91
Warshaw, Robert, 11
Watson, James, 7, 39–40, 198
Wattenberg, Ben, 173
Weinberg, Steven, 39
Weldon, Fay, 177
Welfare, criticism of, 130–132, 184–185
Wertham, Frederick, 134
What You Can Change and What You Can't (Seligman), 145
White supremists, 178–180
Why Women Worry (Handly), 109
Will, George, 115
Wills, Christopher, 8
Wilson, E.O., 39, 106

Wilson, James Q., 115, 133
Winfrey, Ophrah, 84, 92
Women. *See* Gender differences; Mothers
Women's Intuition (Davis), 109
Women Who Run with the Wolves (Estes), 111
Working mothers, 141–142

Wright, Robert, 78
Wrongful birth cases, 160

X-Men comics, 45–46
XYY males, 83–84

Young, Robert, 10
Yount, Rena, 190

Zero family, 25